OBSERVING SOCIETY

OBSERVING SOCIETY

Meaning, Communication, and Social Systems

DANIEL B. LEE AND ACHIM BROSZIEWSKI

CAMBRIA
PRESS
AMHERST, NEW YORK

Printed in the United States of America

Requests for permission should be directed to:
permissions@cambriapress.com, or mailed to:
Cambria Press
20 Northpointe Parkway, Suite 188
Amherst, NY 14228

Library of Congress Cataloging-in-Publication Data

Lee, Daniel B.
Observing society : meaning, communication, and social systems / Daniel B. Lee and Achim Brosziewski.
p. cm.
Includes bibliographical references and index.
ISBN 978-1-60497-639-7 (alk. paper)
1. Social systems. 2. System theory. 3. Sociology. I. Brosziewski, Achim. II. Title.

HM701.L44 2009
301—dc22

2009028654

To the Fulbright Program
and the German Academic Exchange Service
for making our collaboration possible

Table of Contents

List of Figures

OBSERVING SOCIETY

Introduction

Contemporary Social System Theory

Each individual member of a social group processes thoughts and interprets the world in a completely private manner, yet moments of mutual understanding seem to occur on a regular basis. A man shows a ring, and a woman says (with some luck), “yes.” A pistol is displayed, and a wallet is sacrificed. A ball is kicked into a net, and a crowd roars. The ring, pistol, wallet, ball, net, and the men and women acting may all be said to exist in the world, but their reality as objects does not determine what these things will *mean* to different people. Though they may perceive the same objects, people develop a sense for the meaning of things in the world entirely on their own terms. What can explain why two people would ever agree that a ring indicates a proposal, a wallet keeps a pistol quiet, and a ball entering a net produces both triumph and defeat in the same instant? The achievement of shared understanding among different minds is highly improbable and requires a sociological explanation. Members of groups learn to assume that other members will condition themselves through the independent observation of society to associate certain objects and behaviors with certain meanings. Thus, as we suggest in this book, it is society that produces meaning and understanding for its participants.

Sociologists typically explain understanding between people by pointing to a nebulous process known as "socialization." Supposedly, society enters into the minds of its participants, culturing the raw state of their human consciousness. In the natural state, people are mentally disassociated; but the socialization process organizes and unifies their thoughts. In one leading introductory textbook to sociology, for instance, we learn: "Socialization is the process whereby the helpless infant gradually becomes a self-aware, knowledgeable person, skilled in the ways of the culture into which he or she was born" (Giddens, Duneier and Appelbaum 2007:92). Humans are *born into* a culture, as it were, and this common experience slowly transforms them into members of society. Thus, the miracle of societal members understanding each other can be explained by their being born into the same culture. This traditional account of socialization leaves many important questions unanswered. For example, does the infant find itself inside a culture from the moment of its birth, all at once? Can a person leave one culture and be "reborn" into another one? Are people born into two different cultures inherently unable to understand one another? And why is it that people born into the same culture sometimes misunderstand each other?

The concept of socialization implies that people understand one another because the same society is literally "inside" them, or that the people are inside of the same society. The "process of socialization" might initially appear to explain how people learn to understand each other, but only if we accept two rather dissatisfying conditions. First, we must substitute the problem of how society gets "inside" its members for a different problem: how people are born into the same culture. Second, we must overlook the problem that mainstream sociology has not effectively defined society or determined its boundaries. How can people be in or out of a space that has not yet been delineated? Anthony Giddens and his colleagues, explaining to students beginning the study of sociology, say: "A society is a *system of interrelationships* that connects individuals together. No culture could exist without a society; and equally, no society could exist without culture" (2007:59). This definition ties society to a "system of interrelationships" between individuals, and also proposes that society and culture are interdependent. But what is the difference between society and culture? If we accept that individuals are born into the same culture, why do we need a concept of society? If society refers to a system of interrelationships between people, what specific difference does culture make? And exactly how do these different "interrelationships" become part of a single system?

Sociologists have acquired the habit of using terms such as society and culture without defining them in a disciplined and scientific manner. "Once it escapes the everyday world of ethnography and ethnomethodology," Dorothy Smith argued, "sociology's characteristic lexical practices allow it to float free of any actuality capable of constituting a common ground of reference" (2001:165). Smith asserted that it is easy to make fun of the way sociology has developed an appetite for "non-observable" concepts that are assumed to exist without requiring a clear reference or any additional intellectual worries, supported only by what she calls a "blob-ontology." Too often, with their most basic terms in such a muddled state, sociologists have been unable to account for the relationship between individuals and society. The problem of how people use society to manage one another's understanding of objects, actions, and events remains unsolved. We suggest it is time for a dramatic paradigm shift in sociology. Traditionally, sociologists have focused their attention on people, actors, agents, human subjects, and actions. This book departs from that well-worn path and instead turns to contemporary social system theory and an analysis of society comprised of units of *communication*. By "observing society," we mean observing the conditions under which communication accomplishes an understanding that allows for further communication and the reproduction of society as a social system.

AN INTRODUCTION TO CONTEMPORARY SOCIAL SYSTEM THEORY

This book offers a basic introduction to a contemporary theory of social systems, as developed by the German sociologist Niklas Luhmann and other contributors. Social system theory describes society as an autopoietic, self-organizing system that does not include human beings. By using the term *autopoiesis*, a term adapted from the biologist Humberto Maturana (1981), social system theory asserts that society is a closed social system that produces itself and its environment by recursively connecting its own elements, by establishing selective relations between its ongoing communicative operations. As life produces life, society produces society. Society organizes its elements according to its own logic, selectively steering itself into an open future by meaningfully connecting operation after operation. In this book, we explore the multidisciplinary roots of this new paradigm and assess its tremendous potential to redirect sociological reflection.

Contemporary social system theory creatively builds upon the classical theory of general systems, as developed by Ludwig von Bertalanffy (1950, 1968) and imported into sociology by Talcott Parsons and his colleagues (Parsons 1951, 1963; Parsons and Shils 1951). Modern social system theory incorporates fresh insights gained from cognitive biology, the philosophy of consciousness, phenomenology, distinction theory, socio-cybernetics, and constructivism. The theory maintains that observers recursively self-construct everything that is meaningful in the world, beginning with the difference between themselves and their environment. They use this difference to self-referentially process changes in their environment in the form of other-reference, as if information were externally available. Consulting other-reference, an observer is able to make informed decisions while selectively performing operations. The identity of observer may be attributed to individual people, interaction groups, organizations, functionally differentiated systems (such as the economy, politics, law, and religion), and to society itself. Social system theory is concerned with the second-order observation of differences, distinctions, relations, and boundaries—it is not interested in defining essences, laws, values, or ontology. This makes contemporary social system theory very different from both the general systems approach of the 1950's and from other mainstream sociological paradigms.

Within social system theory, individual human observers are viewed as psychic systems, self-conscious unities of thoughts that are operationally closed to everything outside of consciousness. Thoughts remain trapped inside each individual's awareness and cannot make a difference or take the form of information outside of the mind. To participate in society, psychic systems assert themselves as part of the environment of communication. The theory criticizes sociological paradigms that attempt to solve the problems of operational closure and double contingency with presumptions of intersubjectivity, collective conscience, communicative rationality, or normative consensus. There is no inherent order that supports modern society or guarantees understanding among humans. Social system theory emphasizes the improbability of successful communication and yet explains how society still manages to reproduce itself.

In this book, we depict modern society as a complex unity with functionally differentiated parts. Various societal systems (the family, the economy, politics, religion, science, art, and others) produce and organize special forms of differentiated communication that may be distinguished by their own self-constructed boundaries. Each societal system functions in an exclusive way, seeking only to combine its own problems and solutions. Every social operation, however, happens inside the boundary of communication. Social system theory unravels this paradox of unity in difference by depicting society not as a whole composed of parts, but rather as the difference between everything that is communicated and everything that is not. Sociology, from the perspective of social system theory, is the science devoted to describing and understanding the second-order observations of observers of society. Sociologists observe participants in society to understand how participants observe communication processing meaning to recursively index differences in the world, maintain its various boundaries, and construct the information used in its selective operations.

Contemporary social system theory offers a highly original description of modern social conditions and the possibilities of communication.

There is no possible way to scientifically describe the observation of society as a whole because the complexity of functional differentiation removes any operational center from the social world. Every perspective yields useful information only by closing itself, by selecting and limiting its field of view. Variety, restriction, and self-referential closure set contingent parameters for making multiple forms of meaning in a world society that offers no all-encompassing Archimedean vantage point. With the continued autopoiesis of modern communication, society observes its own distinctions in a decentralized and polycontextural manner. The future of communication is open, therefore, because society has closed itself.

This theory of society presents itself as the key to what Niklas Luhmann called a "sociological enlightenment" that will bring social thought out of "the Dark Ages" of anthropocentrism, social ontology, and normative social philosophy. In this book we explain why Luhmann and other social system theorists argued that sociologists have continually failed to produce an adequate theory of society. Orthodox sociologists occasionally use "theory" to help organize research and account for a myriad of empirical data. However, theorists spend most of their time interpreting and reinterpreting "classical" treatises, as if sociologists could understand society by studying their own revered texts. When it comes to theory, traditional sociology is a quagmire of competing grand narratives, none of which offers a generally useful concept of the unity of society. The discipline has assembled a small collection of favorite theories, including conflict theory, functionalism, rational choice, feminism, symbolic interactionism, and others. Students in introductory sociology courses are faithfully instructed in the canonical "paradigms of the masters," an enculturation process that only intensifies in graduate school. With so many divergent theories in use, it can be difficult for students to see any semblance of an integrated approach within the discipline. Indeed, it is often difficult to find consensus even within each of the most popular theory camps.

We believe that sociologists have trouble making sense out of one another's work because they lack a systematic way of describing what they see in the social world. And when sociologists try to explain their observations to non-sociologists, they are often accused of merely wrapping up common sense in an overly complicated package. In this book, we demonstrate that sociologists can use the same concepts from social system theory to describe very different aspects of society. They can communicate their findings using a common theoretical vocabulary, even while describing how the distinctions they employ are limited, contingent, and self-constructed. Supported by a transparent epistemology, the concepts used by social system theorists are compelling because they can be applied universally, to all parts of society, even to sociology itself. Dispensing with the old grand narratives and profiting from recent advances in other fields of inquiry, contemporary social system theory extends the reach and increases the descriptive power of sociological discourse. As a consequence, social system theory promises to increase sociology's plausibility as a science, its research productivity, its connection to other disciplines, and its resonance with the public.

The Structure of This Book

This book is comprised of eight chapters. It begins with an epistemological account of observation that entails the drawing of a distinction and the indication of one side or another. Observing requires one to imagine differences such as system and environment, actual and potential, form and medium, information and utterance, or language and noise. Imagining a distinction sets the stage for an observer to meaningfully indicate or select a preference from a menu of alternatives. Sociologists are already familiar with the nature of scientific observation and with the methodology of participant observation, but our use of the term may come as a surprise. Social system theory takes a very precise approach to the problem of observing and extends this novel perspective throughout its reach. Our initial discussion specifies the influence of George Spencer Brown and relates his logic of observation to the philosophy of Alfred North Whitehead, the constructivism of Heinz von Foerster, and to the sociological work of other relevant thinkers. In this first chapter, we also present definitions of information, meaning, recursion, and culture that will be used throughout the book.

In the second chapter, we define Niklas Luhmann's path-breaking conception of communication as an "improbable" synthesis of information, utterance, and understanding. A focus on communication as the unit of sociological analysis is revolutionary. For readers well versed in traditional sociology, this chapter may be difficult to digest because it ultimately breaks with established patterns of analysis. It also substantially modifies Claude Shannon's and Warren Weaver's famous model of communication, in which information is transmitted between a sender and a receiver (1949). We sincerely hope that any difficulties readers experience will be tied to the intellectual controversy over how sociologists ought to define their subject matter, rather than to our efforts to present the theory in a comprehensible manner. The idea that communication must be observed as a synthesis of three distinct selections is theoretically complex and for this reason we have included many illustrations and practical examples.

Social system theory places the problem of understanding at the heart of every investigation. In chapter 3, we outline the contours of a basic problem: people have different brains, so how is understanding among members of a group possible? To provide a sociological context, from which our theory eventually emerges, we critically evaluate a diversity of theories about how members of society share "intersubjectivity." We also show that pioneering sociologists focused a lot of attention upon the power of communication to cultivate forms of subjectivity and inform social meaning, but their work has failed to establish an adequate and integrated theory of how society produces understanding for its observers.

In chapter 4, we describe language as a medium of communication. Building on the previous discussion of how each observation involves a two-sided distinction, we explain that all communicative operations make use of the binary structure of language. Although it is a highly improbable contrivance, language works because it takes form within the medium of acoustic or optical noise and can therefore serve as a structural coupling between different people. To convey information to an observer, of course, language must also represent more than perceived noise. By participating in verbal communication, participants learn to draw a distinction between organized and unorganized noise. Language is used to code world events and meaning. Furthermore, it doubles all events and all meaning. Everything that can be said, can be said in a positive or in a negative way: in a sentence using "no," or in a sentence using "yes." Language itself is coded using this basic binary structure, a decisive characteristic when it comes to communicative innovation and, in the long run, to social evolution. To clarify the unique qualities of language, we compare and contrast noise-specification in humans and animals and discuss the human voice box from an evolutionary perspective.

Social system theory explains how communication increases its ability to successfully establish different kinds of connections by constructing symbolically generalized media or success media, as we explain in chapter 5. Symbolically generalized media help participants in society assess and manage the contingencies of one another's motivations. The use of symbolically generalized media appears to "cause" the acceptance of communicated propositions. We focus our discussion on four success media: love, power, truth, and money. We conclude the chapter by relating the differentiation of success media to the differentiation of symbiotic mechanisms that structurally couple society and human bodies.

Contemporary social system theory describes society as the overarching system that includes all communication. On the one hand, the all-encompassing unity of society cannot be observed, but it can be delineated clearly in terms of the theory. On the other hand, very small "ephemeral, trivial, short lived" social systems may appear whenever observers rely on the autopoiesis of communication to cope with the experience of double contingency (Luhmann 1997a:812). We distinguish three types of relatively stable social systems in chapter 6, each one drawing a distinction between itself and its environment by means of an exclusive distinction between system and environment. Interactions are observable as the most immediate social systems and include partners in communication who are present for one another. Organizations are observed as social systems comprised of decisions made within a network of members. Finally, societal systems are functionally differentiated channels of communication that pursue operations that refer to a specific problem area of society, such as the economy, law, education, religion, science, and the family.

In chapter 7, we further develop the construct of social differentiation by describing four forms in which society may turn inward and internally differentiate itself: as segmented, center/periphery, stratified, and functionally differentiated communication. In this sense, we compare increasingly complex alternative forms in which society may observe the possibilities for continuing and reproducing its unity. A form of society organizes and selects sequential elements of communication by observing a working distinction, but this distinction may change as time passes, conditions change, and society evolves. Segmented, or tribal, society is the oldest and simplest form, analogous to Emile Durkheim's "mechanical" society (1964), Ferdinand Toennies' "*Gemeinschaft*" (2001), and Robert Redfield's notion of "primitive" society (1953). The form of society increases in internal complexity when communication is differentiated according to the distinction between center and periphery. Stratified society, as Max Weber (1972) nearly suggested with his concept of "status group," organizes communication in terms of position within a stratification scheme, making reference to a socially established hierarchy. Finally, in functionally differentiated modern society, communication is organized according to its function. As it continues to evolve, society may establish other patterns of connections, but these four forms have reproduced themselves with consistency.

As with any other sociological theory, contemporary social system theory must demonstrate an ability to contribute to a method that effectively supports scientific inquiry. In our final chapter we observe social system theory and evaluate its potential to inform empirical research. We argue that the traditional form of the datum (datum/variable and constant) is the fundamental form used in scientific observation. Any useful social system theoretical analysis depends on a method that can specify the form of the datum and describe how its "empirical" variety is conditioned. Qualitative methods such as interviewing, content analysis, discourse analysis, and ethnography have been productively used by social system theorists. However, we outline Luhmann's pioneering functional and comparative method that is uniquely relevant for observing how communication selects and relates its own problems and solutions. Drawing a distinction between problem and solution recreates the familiar two-sided form of the datum: marking variables and their possible actualizations on one side, while indicating constants on the other. Observing society, the quintessential problem system, construct and selectively connect its differentiated problems and solutions necessarily integrates theory and method. The functional method, we believe, has the greatest potential for demonstrating the value of social system theory's resources.

The Attraction of Sociological Theory

We were motivated to write this book by a number of different interests. The incredible work of Niklas Luhmann has yet to be appreciated outside of the German speaking world. Part of the problem is that many of his texts, including his capstone work, *Die Gesellschaft der Gesellschaft* (The Society of Society), have yet to be translated into English. Another part of the problem is the challenge of interpreting his creative and highly abstract ideas. "Complicated conceptual relationships of this kind," Luhmann asserted, "may intimidate sociologists" (1995a:488). Both the authors of the present work have taught social system theory in the United States and Germany and we have seen even our most capable students fall into a state of despair after spending time reading Luhmann. In German, and in English, as within any other tongue, it takes time to learn the language of the theory. Luhmann is the first to admit that his writing is not easy to digest. His use of highly abstract concepts seems to serve as a way of warning readers not to speed along without realizing just how much he wants to adjust the sociological imagination and its resources. He wanted clear and decisive new concepts, new ways to connect ideas, and a new way of disciplining sociology. We try to convey Luhmann's ideas without losing their inventive and sophisticated quality. Our intention is to present the theory to a larger audience, to students in particular, without sacrificing authenticity and intellectual depth. Our strategy incorporates three techniques: we allow Luhmann and other contributors to speak in their own words as often as possible, we consistently name components of the theory so that they may be easily recognized in successive chapters and also traced back to their original sources, and we illustrate difficult concepts with clear and accessible examples. We hope that this book will provide a foundation upon which readers may build, whether they decide to contribute to the theory or search for another way to develop sociological thought.

We believe that social system theory solves important problems that have long confounded sociological theorists. The distinction between microscopic and macroscopic sociological analysis, for instance, has divided theorists into relatively autonomous and mutually resistant camps. Although several influential sociologists have attempted to integrate, synthesize, or form linkages between micro- and macro-level analysis (Alexander et al. 1987), the conceptual tools and methodologies that could support constructive conversations within the discipline have proven elusive so far. Jeffrey Alexander, for example, proposed a general theory that "allows us to link action more closely to meaning as actors themselves experience it and to the cultural forms" that structure meaning (1998:220); but he ended championing a collectivist and normative perspective and rejecting micro- or individual-level analysis (1985). Randall Collins proposed another general theory; but he ended up describing himself as a radical micro-sociologist. All macro-level social phenomena, he asserted, can be traced to interaction ritual chains produced at the micro level (1981; 2005). We could discuss other attempts to integrate micro- and macro-level analysis, but our interest is limited to explaining how social system theorists aim to observe communication at every level of analysis, from the dyad to world society, and from interactions to organizations and functional systems of society. One must always select the social system to observe, but the theory has no inherent preference for micro or macro levels of analysis. Every system is observed at the same level of abstraction and described with the same theoretical resources. In addition, social system theory explains how social structure and agency—a duality that has proven so difficult for sociologists to conceptualize—emerge together in the same instant. There are no communicative structures without actors, and no actors without structures of communication. Society creates "the people" who produce it. The thorny relationship between the individual and society is also neatly resolved by social system theory. Sociology has never been able to explain how participants enter or become part of society. The theory at hand does not try to do so; society is possible *because* participants never become part of it. At this point, we can only hint at how the innovative positions explained in this book will enhance sociological theory, but this will be fully explained in subsequent chapters.

Finally, we were also motivated to share social system theory with readers because we find sociological theory a fascinating, exciting, and worthwhile intellectual adventure. Without even having a definition of society, sociology has made significant contributions to our understanding of the differences resulting from the the division of labor, secularization, urbanization, democracy, and other social phenomena. Contemporary society poses many new challenges for sociology that cannot be met with the old familiar resources. Sociologists have responded in a variety of ways, producing texts that offer moral guidance, journalistic reports, political commentary, narratives of identity, demographic descriptions, and new and improved interpretations of the discipline's founding writers. By contrast, social system theory offers clear concepts, a promising research agenda, and a cutting edge description of modern society. We believe that the theory can support a sociology that can speak to emerging trends in a more powerful, confident, and authoritative voice. Students are irritated (understandably) by sociologists who address endless contemporary phenomena for which they have no words of their own: globalization, digital media, virtual communities, transnational epidemics, cyberterrorism, chemically augmented personalities, cyborgs, the digital divide, information overload, technological risk management, postindustrialism, and many others. Using the resources of social system theory, sociology can investigate these aspects of modernity on its own terms, building its own self-reference, without resorting to poorly imitating the discourse of other disciplines. Sociological investigations can stay centered on communication and reveal how society organizes variety, selects variation, and stabilizes forms of meaning. We are interested in promoting social system theory because we have found no comparable alternative for stimulating and supporting our interest in the observation of contemporary society.

CHAPTER 1

DISTINCTIONS

OBSERVATION, MEANING, AND THE REDUCTION OF COMPLEXITY

In 1969 George Spencer Brown, a British mathematician, published an enigmatic and controversial book called *Laws of Form*. The theme of the book, as described by its author, is that "a universe comes into being when a space is severed or taken apart:"

> The skin of a living organism cuts off an outside from an inside. So does the circumference of a circle in a plane. By tracing the way we represent such a severance, we can begin to reconstruct, with an accuracy and coverage that appear almost uncanny, the basic forms underlying linguistic, mathematical, physical, and biological science, and can begin to see how the familiar laws of our own experience follow inexorably from the original act of severance. The act is itself already remembered, even if unconsciously, as our first attempt to distinguish different things in a world where, in the first place, the boundaries can be drawn anywhere we please. (Spencer Brown 1979: v)

According to Spencer Brown's innovative "calculus of indications," observing any form of difference always entails two steps: drawing a two-sided distinction and, in the same operation, selecting or indicating one side or the other.

Although his work received mixed reviews from his own peers, Spencer Brown's impact on the German sociologist Niklas Luhmann was critical. One of the most prominent, creative, and prolific sociologists of the twentieth century, Luhmann acknowledged the method of observing through distinctions as a guiding principle in his scholarship. The pioneer of contemporary social system theory cited no more than a few selected pages from *Laws of Form*, but Luhmann repeatedly returned to the problem of how an observer draws a distinction to produce a world. For instance, in his book, *Art as a Social System*, Luhmann wrote,

> Since observation depends on distinction, it must dissolve and make invisible the underlying paradox of the unity of the distinguishing activity, it must replace this unity with operatively useful distinctions, it must unfold the paradox—there is no other way of arriving at identities capable of operation.
>
> The first caesura, the first cut into the unmarked state of the world, must be executed, and not just to generate two sides. Rather, the cut must yield an asymmetrical use of these sides, which allows for connecting operations on one side but not on the other. This is how sequences are set in motion that reproduce the problem in the realm of already executed distinctions in such a way that further observations can follow. (2000a:42)

The first cut made by an observer of society divides communication from its environment, from all that is not communication. Society *is* communication, as we will make clear, and it is the superset that contains all social systems. An observer of a social system identifies unities of sequential communicative operations, operations that successfully establish connectivity, one observation at a time, with the emerging history of society. This technique of observing with reference to a clear distinction shapes the way in which social system theory redefines the most significant sociological concepts, including culture, symbol, meaning, and structure.

This book provides a comprehensive introduction to social system theory and explains how its revolutionary method of observing society as communication is making a major contribution to modern sociology. Because of the fundamental importance of observation in social system theory, we devote our first chapter to clarifying the difference-oriented epistemology Niklas Luhmann developed from his reading of Spencer Brown. Each of the following chapters will return to the basic premise that sociology observes observers of society drawing cultured distinctions and using them to produce and reproduce slices of the social world in ways that can be anticipated.

Distinctions, Symbols, and Observers

Distinctions

Social system theory develops its underlying epistemology by combining Spencer Brown's methodology of observation with constructivism, an approach emphasizing the experiential origin and problem-solving nature of knowledge, as developed by Heinz von Foerster, Ernst von Glazersfeld, Fritz Heider, and others. In the discussion below, we relate the theories of observation and constructivism to similar concepts developed by a number of influential thinkers, including Ludwig Wittgenstein and Alfred North Whitehead, whose work on self-reference, symbols, and meaning preceded social system theory.

We asserted previously that the consequences of every thoughtful inquiry depend on an observer's "first cut," on the initial decision that opens up a contingent universe of potential connections by drawing boundaries that did not necessarily have to be drawn. Conscious thought is always intentional, as the phenomenologist Edmund Husserl maintained (1982), and every intention requires a distinction to mark it against an overwhelming surplus of alternative possibilities. Niklas Luhmann, directly citing Husserl, carefully connects his own pivotal concept of meaning to the notion of an observer's intentionality:

> [Systems] produce an enormous number of surplus possibilities that appear for an observer (which may be the mind itself, another mind or a social system) as *meaning*. Meaning is nothing but self-imputed selectivity, based on the construction of a vast amount of connective possibilities. At the phenomenological level, meaning appears as a 'horizon' of further possibilities, co-presented with every focus of actual attention or communicative thematization. (Luhmann 1996c:343)

As defined by Luhmann, meaning is the difference, constructed by an observer, between the actual and the potential. The recursive process of infusing an "actual" observation with meaning by associating it with past observations involves the operation of a "memory function" (Luhmann 1997b). According to Heinz von Foerster (1987), cognition gains "depth" or complexity as sequentially indicated distinctions are *recursively* connected and placed into meaningful relationships. If an observer could not remember (and forget) the meaning of distinctions drawn in the past, its own self-reference and internal complexity would overwhelm itself and could not evolve or be enhanced by culture. Without its past, traceable all the way back to its first distinction, self-reference has no present and no future. An observer consults memory to intentionally relate or connect two or more different observations to each other, self-referentially using meaning to produce unity through generalization and comparison (Luhmann 1995a:95, 1997a:588). Meaning "renders complexity" through a difference-oriented processing of information, characterized by a constant shifting of actuality (Luhmann 1995a:61–63).

To name the imagined unity that is produced when distinct observations are connected with meaning, we use the term "system." Psychic and social systems are treated as observers who draw distinctions and make indications using self-referential forms of meaning. In this sense, psychic and social systems may be considered separate but parallel processors of meaning, both of which are essential in the production of society. Subsequent chapters will further differentiate and relate both types of meaning-based systems in much more detail, but this chapter concentrates on explaining the general concept of observation.

Psychic and social systems are observers that operate by recursively connecting their own observations. Psychic systems handle distinctions within the medium of consciousness, but social systems, as Luhmann asserted, "must acquire their own possibilities of observation" (Luhmann 1995: 36–37). Without the ability to construct possible connections between successive observations, a meaning based system would confront only chaos resulting from the world's overwhelming variety. Put in stronger terms, a psychic system that could not observe a distinction between "world" and "everything else," would not be able to perceive "a world" and establish connections between phenomena attributed to its world. Observations of difference make identification possible. In the words of the philosopher Alfred North Whitehead, "Synthesis and analysis require each other" (1959:26). Things are what they are because they refer to *other* things. That is, things are what they are because of what they are not. Meaningful observations construct and refer to this difference.

Before we expand on the connection between making observations and social system theory, let us briefly return to Spencer Brown's notion of "severance" and his own illustration of the theory of observation with reference to the circumference of a circle in a plane (Spencer Brown 1979). Imagine drawing a black circle on a white piece of paper with the help of a compass:

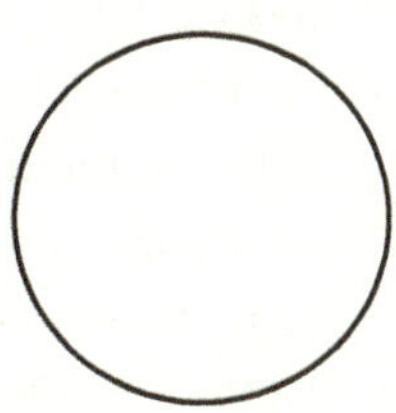

What makes the circle a circle? The circle we see is not formed by the space *inside* the line, the space *outside* the line, or by the line itself. Identifying a circle requires a synthesizing unit of meaning that consolidates all three of these components while also preserving their distinctiveness, a paradoxical unity in difference that represents what social systems theory recognizes as a "form" (Luhmann 2001a:257). The distinction or severance between the form's inside and outside is produced by the drawn line because that is the boundary between them. There is virtually no limit to what is included on the outside of the circle. To observe the circle we indicate or mark the space inside the line, but we could just as easily "cross over" and indicate the outside, which might yield the appearance of a paper with a hole in it, a rusty screw on the compass, or a dog lying in the corner of the room. Our self-referential calling determines whether we indicate the inside or the outside, but observing the circle requires drawing a distinction between the two sides. Every observation (and every system) entails a marked space, an unmarked space, and a self-referentially constructed border that produces the difference. The two-sided unity in difference or "the form" with which an observer continually operates may become the subject of self-reflection and critique because it was itself selected among alternative forms (see Baecker 1999). Once again, both the form and the particular side of that form chosen result from an observer's decisions; they are two different choices, both of which reveal the self-reference of the observing system.

Observations make sense because of self-reference. Every observer—by drawing a distinction and indicating one side—does so according to their own logic. Working alone, an observing social or psychic system produces its own account of why it selects one side of a distinction and not the other. While it cannot choose both sides without trapping itself in a loop of uncertainty, the observing system can pick one side in this moment and switch to the other side in the next, oscillating between the two possibilities of the form. An indication and a crossing cannot happen simultaneously—each captures its own moment in time. Nonetheless, whenever the observing system has a reason to select one side and not the other, it is certain that the meaning behind the choice was generated *inside* the system itself. This point will eventually take on fundamental importance when we connect it to the binary codes and programs used by social systems. Codes set up distinctions, but programs inform observers how to make meaningful choices presented by the forms of meaning used in their operations.

Symbols

For a psychic system, meaning may, for example, be recursively indexed by a string wrapped around a wrist. To the conscious mind, the string might be a sign that indicates the imaginary message: "remember to email your aunt and wish her a happy birthday." The meaningful connection between the string and the aunt is self-referential and depends on a psychic system's own potential to use signs to help recall memories. One sees the string and remembers, hopefully, an aunt who ought to be e-mailed. If the string "works," rendering the intended connection, its meaning is understood. If the message is received and understood, it is not because the string is tied to a particular message. In fact, the string might just as easily be used to help convey a completely different message: "remember to buy milk on the way home." The actual meaning of the string can be observed in opposition to other possible meanings only by the person who wrapped it around their wrist. The string could potentially mean anything else besides, "e-mail your aunt." The indicated side of the distinction is differentiated through self-reference. Without the meaning supplied by self-reference, the string is just a string and the aunt may be forgotten.

In 1927 Alfred North Whitehead presented a series of lectures at the University of Virginia on the meaning and effect of "symbolism." His ideas foreshadowed the way social system theorists now conceive of meaning and language. Whitehead's formal definition of symbolism included both the notion of distinction and of self-reference:

> The human mind is functioning symbolically when some components of its experience elicit consciousness, beliefs, emotions, and usages, respecting other components of its experience. The former set of components are the "symbols," and the latter set constitute the "meaning" of the symbols. The organic functioning whereby there is transition from the symbol to the meaning will be called "symbolic reference." (1959:7–8)

In Whitehead's terms, individual observers perceive things in the world using their senses. Instead of seeing each distinct object as a new and unique phenomenon, humans creatively analyze perceived differences and actively synthesize meaningful symbolic unities. The mind perceives its sensations as "real." However, in order to make meaning of its scattered impressions, the mind uses symbolic generalizations to construct its own sense of reality that is filled with imaginary redundancies. Thus, symbols and signs are linked to meanings only in the minds of observers.

Observing the meaning of symbols is a self-referential process. As Whitehead argued: "We must conceive of perception in the light of a primary phase in the self-production of an occasion of actual existence... the total activity involved in perception of the symbolic reference must be referred to the percipient" (1959:8). Perception is the product of an "internal relationship" between the perceiver and the thing perceived. According to Ernst Cassirer, symbols "mediate" between perception and understanding (1965). The world is not accessible in a direct and unequivocal manner. To make any difference for an observer, impressions of space, time, and matter must be symbolically processed. What is perceived can be given symbolic meaning by an observer who connects and enriches perceptions with self-reference. According to Niklas Luhmann, cognition "must be understood as a recursive processing of symbols in systems isolated by the conditions of the connectibility of their own operations" (2002:171).

The imaginary association between a symbol and a meaning is a mere contingency that could just as easily have taken on a different form. Whitehead emphasizes that "A symbol and its meaning do not require *either* that there shall be a symbolic reference between the two, *or* that the symbolic reference between the members of the couple should be one way on rather than the other way on" (1959:10). Whitehead poses the following question: "Why do we say that the word 'tree'—spoken or written—is a symbol to us for trees? Both the word itself and trees themselves enter into our experience on equal terms; and it would be just as sensible, viewing the question abstractedly, for trees to symbolize the word 'tree' as for the word to symbolize the trees" (1959:12).

Whitehead's query about why the word-symbol for "tree" should indicate a tree (and not the other way around) is stirring because it evokes the paradox of an unresolved binary distinction. There is no a priori information or ontology available to determine what should be indicated or selected. The paradox can only be resolved when an observing system indicates one or the other side according to its own self-constructed information. When it comes to attaching symbols and objects in the world, there is the possibility to observe society, and to inform and constrain one's own use of symbols with society's references. It takes an observer to perceive a tree (to see the object standing before them). Perceiving others also confronted by the tree, an observer may come to expect that others will identify the tree as such, and also expect others will share this same expectation in return. The reciprocity of this expectation depends, however, on the self-reference of society providing names for the objects perceived by observers. Without society's interference, an observer could not call a tree by any name.

Observers

As the sociologist Stephan Fuchs maintained, the world contains no information, only unstructured complexity. Any information that makes a difference for an observer is the self-made product of the observer (Fuchs 2001:17–18). Thus, every whimsical observation made by an observer, every self-referential link between a symbol and a meaning, can provide information about that observer. Fuchs pointed out,

> Empirical observers observe the world in this world, not from outside. As part of the world, observing and observers emerge and disappear at certain places and times. Their empirical status implies that they can also be observed themselves, by other empirical observers or by an observer who observes himself. (2001:18)

Information about an observer may be used reflexively by the same system, with self-observation leading to "self-understanding." Such information might also be of interest to another observing system, leading to "knowledge of the other." "Observations," Fuchs suggested, "always tell at least as much about the observer as about the referent of observation" (2001:24).

Understanding how observers make meaningful connections between operations is, in Luhmann's words, the "pre-eminent question" for a theory of self-producing systems:

> How does one get from one elemental event to the next? Here, the basic problem lies not in *repetition* but in *connectivity*. The differentiation of a self-referentially closed network of reproduction proves to be indispensable exactly in view of this problem of connectivity; and it is possible to formulate problems of the formation and change of structure only in respect to a system formed by such a network. It is structures, in other words, that must make possible the connectivity of autopoietic reproduction if they do not want to give up the basis for their own existence, and this limits the domain of possible changes, of possible learning. (1995a:36)

The work of Whitehead, Cassirer, and others steers us to an awareness of how observers construct their world through symbols. Under the influence of George Spencer Brown, Niklas Luhmann and Stephan Fuchs lead us to appreciate what can be learned by observing how different observers process the meaning of their own observations by making symbolic, imaginary connections between them. For social system theory, the main subject of interest is not the psychic system of the individual; but society as an observer of its own kind. How does society use symbols to mediate its reality, a reality that may consequently inform the observations of multiple psychic systems?

Contemporary social system theory seeks to describe how social systems draw distinctions between themselves and their environments, how social systems self-referentially organize the complexity of their worlds with the help of symbolic forms, and how social systems reproduce themselves by remembering and connecting meaningful observations (Luhmann 2006). The central task of sociology is to describe how and under what conditions *particular* observers make particular selections with reference to particular distinctions. In other words, sociologists are second-order observers who trace the associations of meaning that have been selectively observed by first-order observers (in particular, sociologists observing social systems). An observer self-referentially connects observations to build an imaginary system that can guide the continued use of distinctions. Such a system produces itself from its own elements; it is an autopoietic and self-organized creation of an observer. From this perspective, the world is not filled with objects that are simply available for observation. The sociologist observes a first-order observer constructing a world by relating successive observations of its own design. The sociological task is to understand how distinctions are selectively handled by observers of a social system, and to describe any kind of socially stablized objects that might emerge when these observations are recursively connected back to their own results (Luhmann 2001b:270). These associations, however, can only be pursued by a second-order observer observing first-order observers. This limitation represents a blind spot that results from the fact that an observing system cannot simultaneously observe and reflect on the form of the observation. Stephan Fuchs' explained,

> At the first level, one can see something, but not the seeing itself. Optical illusion, for example, can be revealed only later, and only on a different level, which has the first level as its theme. While they observe at this level, first-level observers cannot observe just how they manage to observe. They cannot see the physical, physiological, or social conditions of observing. They also cannot observe just how selective their pattern of observation is, as an actualization of one possibility among alternatives. This becomes an option only at subsequent levels, which have their own conditions for observing. (2001:25)

Depending on the distinction drawn, one can reflexively observe oneself as an observer on the second-order, producing functional, ethical, psychological, sociological, or other kinds of observations about how one operates on the first order. Using the resources of social system theory, therefore, sociology can observe systems as first-order observers, and then observe and describe how sociology is able to make its own observations of systems.

Reducing Complexity: The Function of Memory and Culture

Complexity and Consciousness

When the philosopher David Hume described his own inner-sense of consciousness, he emphasized the significance of each subject of perception:

> For my part, when I enter most intimately into what I call myself, I always stumble on some particular perception or other, of heat or cold, light or shade, love or hatred, pain or pleasure. I never can catch myself at any time without a perception, and never can observe any thing but the perception. When my perceptions are removed for any time, as by sound sleep, so long am I insensible of myself, and may truly be said not to exist. (Hume 1978:525)

In William Luijpen's apt formulation, "If the 'something' which is not consciousness itself is thought away, consciousness itself also is thought away. For otherwise consciousness would have to be thought of as consciousness of nothing, which is not consciousness" (1960:20). Sleep permits the loss of consciousness, which is otherwise consumed with actively pursuing objects. But confronted with so many possible objects, how does consciousness select the "something" with which it occupies itself? Surrounded by a whole world full of things to observe, how do first- and second-order observers select their targets for observation? The world appears to be full of raw and overwhelming complexity. As an observing psychic system, human consciousness must develop techniques for meaningfully selecting its targets from a seemingly endless supply of alternatives. As only one of the countless possible things upon which you could focus your attention at any given instant, the thing you actually select is brought forth by a process of exclusion. Each fleeting thought takes center stage in your consciousness only once, as a separate event in time, before vanishing away forever. In one second, you may be thinking about the meaning of this text. In the next second, your attention may be centered on the time of day, the price of gasoline, or the weather. Consciousness must cope with the never-ceasing departure of its own components. How can anyone retain a thought? Which thought should be remembered and which forgotten? We will eventually redirect this concern for how psychic systems selectively restrict the complexity of consciousness and shift the focus toward society's processing of its communicative operations, which is the primary interest of social system theory.

Consciousness functions as a self-referential system that recursively connects present thoughts to past thoughts, forming unities in difference. In this self-producing fashion, thoughts produce more thoughts, one after another. Before an observation exits the stage of one's immediate awareness, one may (or may not) mark its occurrence with reference to a previously established theme or category of related observations. As a conscious observer, for instance, one may "see" that a white dog and a black dog are both dogs (likely a familiar, pre-established category). White and black are both colors. Dogs bite and so can people. Today and yesterday are both elements of time. Like so, a single thought, observation, sign, or arbitrary symbol can swallow up many others by providing a symbolically generalized "form." Nonetheless, dogs, colors, things that bite, elements of time, and all other things in the world, have no form until they are shaped and structured by an observer. A form, considering its archaic sense, implies the "essence" and "beauty" of an object, in contrast to the "unshapely" matter or physical material of which it is comprised. Only with the help of a self-referential form, can an observer shape and "make sense of" a sensate impression. Social system theory emphasizes the capacity of forms to present two sides, either one of which can be selected to produce meaning with reference to a specifically coded difference.

Observers are compelled to observe in order to survive. In the most basic and vital manner, observation serves an organism by indicating a meaningful difference between the fight or flight response. How does an organism determine whether or not it ought to run away from a possibly dangerous situation? How does it know when to stand and fight, or, in our mostly civilized lives, to stand and act? All contemplative, intellectual thought occurs in respite between these two poles of action and withdrawal. In the absence of any impulse to either run away or face the world, an organism's capacity for observation would be paralyzed by indifference, disinterest, and apathy. In a hostile world, disinterested organisms perish. To survive, therefore, an organism must possess an active and continual interest in observing and settling differences, time after time. Constructivists use the term "viability" to indicate this fundamental interest within an observer. In the course of experiencing conditions in life, an observer will construct problems that appear to need solutions. For example, "should I stay or should I go?" The viability of a solution is also a construction, but it is a construction that helps an observer solve problems and remove obstacles, or else it is not viable (von Glasersfeld 1992:30). When it comes to staying alive and pursuing happiness, psychic systems share many of the same problems. Constructivists draw attention to the fact that observers may observe and learn from one another's attempts to remain viable. By observing others, we may corroborate and inform our personal constructions of problems and solutions. Turning to cultural resources, observers may increase their ability to make useful observations that solve problems and enhance their efforts to remain alive.

Memory

Peter Berger and Thomas Luckmann suggested that individuals construct their sense of reality from a "social stock of knowledge" attained through communication with others. One's private stock of knowledge is a mental storehouse of information available for possible recall. On the social level, members of groups tend to share a "common" stock of knowledge, containing information available to all members. Just as no store can possibly stock every product, one's own stock of knowledge is never all-inclusive. As Berger and Luckmann described it,

> Within the (linguistically circumscribed) semantic fields thus built up it is possible for both biographical and historical experience to be objectified, retained, and accumulated. The accumulation, of course, is selective, with the semantic fields determining what will be retained and what "forgotten" of the total experience of

> both individual and society. By virtue of this accumulation a social stock of knowledge is constituted, which is transmitted from generation to generation and which is available to the individual in everyday life. (1966:40)

Potential information is excluded from one's personal stock of knowledge when it appears to lack relevance for future situations. Why learn to speak Mandarin Chinese if one never plans to use it as a language? Why learn how a computer works as long as it does what you want? Why count how many times vowels were used on the preceding page? Fortunately, healthy psychic systems usually possess "filtering" mechanisms that help eliminate over-stimulation of the senses. Urbanites, for example, as Georg Simmel pointed out, tend to develop a "blasé attitude" that helps shield them from the overwhelming sights, sounds, smells, and social pressures of urban life. In his words, "The essence of the blasé attitude is an indifference to the distinctions between things" (1971:329). A group's stock of knowledge, Erving Goffman noted, may even include specific rules of etiquette requiring members to demonstrate "tactful inattention;" guidelines for politely acting "in an uninterested, uninvolved, unperceiving fashion" (1959:230).

As long as one's memory works as well as any average stock clerk, bits of information may be selectively recalled for "indexed recursion." An index is something that points to something else. In mathematics, an index is a symbol that indicates an operation to be performed or the function or position of a certain value within an ordered environment. Formulas for indexed recursion may be the result, among other things, of a person's tastes, history, education, occupational specialty, and membership in certain groups. In any case, whenever psychic systems select particular observations for inclusion into their personal, self-organized stock of knowledge, they consult their memory to reference observations and to establish meaningful recursive connections. Selecting certain observations for memory, of course, entails the necessary selection of others for forgetting. Forgetting "superfluous" information prevents the system from blocking itself with the residue of its own past observations (Luhmann 1997a:579). Memory can only operate in the present, but it works by referencing the past in preparation for making meaning in the future. "Memory," Luhmann noted, "retrospectively tests the consistency of all current operations with what the system constructs as its reality" (Luhmann 1997a:578–579). The flow of time remains irreversible and the precise moment of each observation passes. But memory is recursive and (moment by moment and observation by observation) it builds up its capacity for extending meaning and increasing its complexity.

Culture

We are accustomed to thinking about the importance of memory for consciousness and the psychic system, but social system theory also describes the performance of social memory, which references meaning within communication. Social memory is created by society in the course of connecting ongoing communicative operations. Every single communication, as we shall explain in the next chapter, observes meaning by referencing previous communication; it cannot stand alone and must be anchored in the memory of past communication. According to Ludwig Wittgenstein, "Every proposition must already have a sense: it cannot be given a sense by affirmation. Indeed its sense is just what is affirmed. And the same applies to negation, etc." (cited in Luhmann 1997a:584). Social memory requires the participation of psychic systems, but it is independent of what particular individuals remember and how they refresh their own memories with the help of communication. From Luhmann's perspective, "Every society [*Gesellschaft*] relies on its own, self-produced memory to assist it in all of its ongoing operations; no society could make its own operations dependent (and how then, except through communication?) on that which is neurophysiologically and psychically given to participants as something known, familiar, or as a fact of the past" (1997a:584–585).

Social system theory uses the term social memory to more effectively describe what social scientists have traditionally meant by "culture." Luhmann argued that,

> Culture is nothing more than the memory of a society; the filter of forgetting/remembering and the utilization of the past to set

> parameters on variety in the future… Culture does not understand itself as the best of all possibilities, but it rather directs the possibilities for comparison and thus also obstructs the view of other possibilities. In other words, culture hinders thought about what one might do instead of what is traditional. The invention of a special concept of culture may be attributed to a situation in which society has become so complex—in which it must forget more and remember more and must reflect on this—that it needs a sorting mechanism to meet these demands. (1997a:588)

Social memory, actualized in communication, facilitates understanding by restricting participants in communication to socially meaningful themes, binding up meaning in society's own self-reference and the possibility of making recognizable connections between past and present. Social memory encourages cultured observers to forget about distinctions that might also have been drawn, and to adopt those two-sided forms of meaning that have worked to solve problems in the past.

The capacity to "have culture," as we understand it, may be attributed to both individual psychic and social systems. Both types of meaning-making systems consult culture so that their observations (thoughts or elements of communication) can be referenced in connection with familiar themes and recursive indices. According to Luhmann, culture is "a supply of possible themes that is available for quick and readily understandable reception" in concrete meaning-making processes:

> We would like to call this supply of themes *culture*, and, if it is reserved specifically for the purposes of communication, *semantics*. Thus an earnest, conservable semantics is a part of culture, namely, of what is handed down to us by the history of concepts and ideas. Culture is not necessarily a normative content for meanings; perhaps it is more like a limitation of meaning (reduction) that makes it possible to distinguish appropriate from inappropriate contributions or even correct from incorrect uses of themes in theme-related communication. (1995a:163)

Culture is a set of formulas for relating contingencies. Social memory cultivates unlikely connections by providing observers with the ability to reciprocally anticipate the use of established, familiar distinctions or forms of meaning.

CULTURED OBSERVATION

We have emphasized that cognition always involves an observation. Every observation requires a distinction and an indication, an actualization of one selection in opposition to all other possibilities. Observers use self-made, self-referential, recursive forms to *inform* their selections with meaning. The unorganized complexity of the world, its inherent indifference and uncertainty, threatens observers with paralysis. Facing an overwhelming world of possibilities, observers consult memory and culture for the difference that can make a meaningful distinction between alternatives. "An observer is always a cultured observer," Stephan Fuchs concluded, "or no observer at all" (2001:339).

Both psychic systems and social systems operate with a memory function that references meaning. To survive as living observers, people need to share in communication with others, letting consciousness cultivate itself with social memory. Reciprocally, communication cannot happen without the participation of people. Nonetheless, as we will soon explain, it is a unity that does not include humans. As Luhmann maintained,

> Psychic and social systems have evolved together. At any time the one kind of system is the necessary environment of the other. This necessity is grounded in the evolution that makes these kinds of systems possible. Persons cannot emerge and continue to exist without social systems, nor can social systems without persons. This co-evolution has led to a common achievement, employed by psychic as well as social systems. Both kinds of systems are ordered according to it, and for both it is binding as the indispensable, undeniable form of their complexity and self-reference. We call this evolutionary achievement "meaning." (Luhmann 1995a:59)

Social system theory's concern for observing how psychic and social systems operate as observers of meaning is founded in the awareness that both survive by evolving together. Nature has made human beings "poor in instincts" and "open to the world" of available possibilities (Gehlen 1978; Plessner 2003). According to sociologists of knowledge Berger and Luckmann,

> The developing human being not only interrelates with a particular natural environment, but with a specific cultural and social order, which is mediated to him by the significant others who have charge of him. Not only is the survival of the human infant dependent upon certain social relationships, the direction of his organismic development is socially determined. From the moment of birth, man's organismic development, and indeed a large part of his biological being as such, are subjected to continuing socially determined interference. (1966:48)

Thanks to "socially determined interference," psychic systems learn to culture themselves as they observe communication. Participation in society distributes the structural and semantic resources psychic systems require in order to organize the complexity of their own observations.

Compared with most other living organisms, human infants are weak and vulnerable. Even if given every material prerequisite for life (food, water, shelter), it would be difficult to imagine how a socially isolated child could draw any distinctions and make indications: seeking out a mate, caring for children, building a home, and observing its own thoughts. This fact becomes self-evident if one considers the ultimate importance of participating in communication for psychic systems. Nature may produce humans, but only the observation of society produces people. People cannot enter or make up social systems as human organisms, as ontological entities; they participate in society only by taking up an address offered by communication. This awareness appears in the pioneering work of George Herbert Mead and other symbolic interactionists, as we explain in chapter 3.

In the preceding pages, we described how social system theory supplements a methodology of observing observers with insights from constructivism, centering its analysis on the processing of meaning with symbols. As we make clear in the following chapters, social system theory uses this innovative technique of observing with two-sided distinctions to analyze how different functional systems of communication operate. The economy, politics, family, art, education, and other societal systems each draw a particular distinction with which to make observations, each one repeatedly using its own binary code to generate, process, and establish connectivity between communicative operations. The theory also uses its account of observing to draw a distinction between medium and form, a critical move toward explaining the relationship between what an observing system can perceive—the environmental changes or "other reference" it can detect—and what it can recognize as self-referential information. The theory of observation establishes the methodology used to describe different types of social systems, along with the unities of communication that emerge as they connect their distinctive observations. Interactions, organizations, and society's various functional systems are differentiated according to the unique distinctions each employs to draw their own operational boundaries, indicate participants, establish connections within society, and increase the complexity of their self-reference.

By opening up its own universe via drawing a distinction between society and its environment, social system theory has prepared itself with the basal self-reference necessary to observe the forward motion of sequential operations reproducing and connecting to this first cut. Social system theory recognizes what it will exclude (humans, gravity, temperature, dreams, etc.) and indicates the preferred side of its form: communication. The social system theorist operates by using the theory to identify and describe other-reference supplied by social systems. Each first-order observer independently draws distinctions and creates forms of meaning. To be successful, sociology must explain how participation in socially organized observation leads psychic systems to use and expect others to use the same socially conditioned forms of meaning. Communication, the focus of our next chapter, organizes conscious observers with constraints produced by its own self-reference.

CHAPTER 2

COMMUNICATION

THE SYNTHESIS OF INFORMATION, UTTERANCE, AND UNDERSTANDING

To understand this text you are reading now, you must observe significant differences between the symbols that make up written language. You pass your attention from one sentence to the next, moving along past "letters" and "words." Notice how your attention does not get hung up on a single letter, syllable, word, or punctuation mark. Readers learn to observe each element of text just long enough to relate it meaningfully to more complex elements within a self-referentially constructed semantic background. As a reader, one's awareness continually oscillates between observing the meaning of writing and observing the meaning of any information that might be extracted from writing. You may soon turn to the next page and read some more, heaping up additional observations and relating them to what came before. If you forget something important, you can flip back through the pages you have already read and perhaps rediscover what you have forgotten. If you manage to make some sense out of this text today, does that mean that you could have also done so yesterday? If someone had given you this book when you were five years old, what would you have been able to understand about it?

Consider yourself as a developing psychic system, increasing in your ability to self-reference and associate observations. As a five year old, you might have had the ability to observe that this is a "book" and not "something else." It would not have been your pet dog, your mother's shoes, or the door to your room. Most five year olds, of course, already have assembled enough information to help them identify a book against a background of countless other things. As a five year old, you might have observed that this was a "book for adults without colorful pictures" and, therefore, not worth further observation. Finding yourself reading this book today, it is safe to say that your ability to connect observations has drastically increased in complexity. You have become a reader by learning how to look for meaning inside of ink ciphers, highly contingent marks on paper produced out of the difference between black and white.

Even if people must mentally and physically struggle to learn how to read and write, those who master the art of text-based communication are likely to appreciate their hard won skills. If you could not observe printed language, your experience with communication might be limited to hearing another person talk. This would mean that your conversation partner would always need to be within earshot. Telephones would give you some freedom from the need to be near your conversation partner, but still you would be limited to talking to your friend whenever you and she were both willing and able to use a phone. Using telephones would also require that you learn to observe and meaningfully connect numbers, companies, bills, answering machines, the normal turn-taking routines that organize conversations, and other contrivances. By contrast, consider how easy it would be to print a text and distribute a hundred different copies. One person easily could reach a hundred people with the same text, provided each knew how to understand and inform himself or herself with what had been written. It would be much more difficult, however, to give a hundred people the same message verbally. You would either have to hunt down each person independently, taking care to always repeat the same word noises, or else capture the entire audience in the same location and time. If you were to send a message as text, many people could read it at their leisure. Furthermore, while dead writers can no longer talk, their written works still can be read. Learning to read and write pays off because it increases one's ability to share observations with other people. This point leads us to the fundamental importance of systems of communication.

As observers, psychic systems value the medium of spoken and written language because human brains cannot communicate. Nature has placed each one of us in a state of mental isolation that cannot be overcome (unless we were to learn to "read" minds, a proposition entertained and politely dismissed in the next chapter). It is as if we were locked inside solitary prison cells, with no chance of seeing or hearing other people. How could we ever get a message through to the outside? The actual physical co-presence of others, even our most intimate companions, cannot change the fact of our mental isolation. As psychic systems we can think, but our thoughts cannot communicate. A system of communication cannot think, but, thanks to self-reference and the self-produced motivations of participants, it can bring forth a flow of captivating signs and symbols. We can "attach" thoughts—our own and those of others—to communication media such as language and text, thereby maintaining the illusion that we are able to reach other minds.

When it operates successfully, a communication system presents an observation, indicating one side of a distinction. On the indicated side, we have an "actuality." On the other side, we have "possibilities." In other words, when we communicate something, we select "it" and not "something else." We present our selection to a conversation partner, and hope that they will be able and willing to make a subsequent observation in relation to our own. Conversations between people might be seen as emerging strings of connected observations, providing their own context and meaning. Luhmann explained,

> When one communicative action follows another, it tests whether the preceding communication was understood. However surprising

> the connecting communication may turn out to be, it is also used to indicate and to observe how it rests on an understanding of the preceding communication. The test can turn out negative, and then it often provides an occasion for reflexive communication about communication. (1995a:143)

If your partner understands a message you sent, it is because you both are accustomed to observing and participating in the same system of communication. That system, society itself, preceded the present understanding between you and your partner, and it created the chance for you to reach one another in a readable or meaningful manner. To those who are willing to observe its selections and stay within its boundaries, society also offers resources to cope with confusion, ambiguity, and disagreement. In the previous chapter we emphasized that every observation entails the drawing of a distinction and the marking of one side or another. We now center our attention on communication, its memory, and its selectivity. In this chapter, we explain what we mean by the concept of communication. We also describe how participants in society contribute to its autopoiesis, even as they attempt to ensure their own viability.

Communication as a Unit of Sociological Analysis

Sociology, according to social system theory, is the study of communication. What is communicated and what is not? What can be used to form information and what cannot? When does communication lead to further communication and when does it simply stop?

It may seem peculiar that sociologists should entertain such questions, because they have traditionally chosen to communicate about individual actors, the interactions of those actors, and the supposedly shared values or ideas of those actors. Max Weber, for instance, focused his attention on explaining and understanding "social actions." Emile Durkheim suggested that social solidarity depends on the existence of a "collective conscience." Karl Marx, pressing the metaphysical limits of his dialectical materialism, argued that individuals inherently possess a common "class consciousness," even though they may be "alienated" from it.

Talcott Parsons and Edward Shils proposed a theory of social systems in which patterns of "normative consensus" (a common orientation toward values or a general "ethos") were "singled out as the most crucial cultural elements in the organization of systems of action" (Parsons and Shils 1951:159, 190). Juergen Habermas (1984) asserted that social systems may stabilize forms of interaction, but social integration can only be produced in the unsystematized "lifeworld;" where true companions seek cooperation in the affairs of everyday life and where "undistorted communication" and an "a priori" commitment to human solidarity prevail.

In contrast to this sociological tradition, contemporary social systems theorists have rejected the idea that members of a social group can share the same thoughts, sentiments, values, or beliefs. As Peter Fuchs noted, "No thought or emotion has ever transcended the barrier of the human skull" (Fuchs 2002a). They also have rejected the assumption that "social acts" can be identified, explained, or understood outside of a theory of communication. During the late 1970s and early 1980s, Luhmann frequently spoke to his students about how difficult it was to resolve the "major theoretical decision" of whether to identify actions or communications as the "best candidate for element status" (Stichweh 2000:7). In Luhmann's path breaking final determination, "only communication can communicate and what we understand as action can be generated only in such a network of communication" (Luhmann 2002:156).

When sociologists attempt to use "social action" as their unit of analysis, they fall into a bottomless analytical pit. How would it be possible to know when an observed action begins and ends? Alfred Schutz argued that sociologists cannot assume that a social action has any kind of analytical unity or defined parameters. In his opinion, when a sociologist describes an observed social action and interprets the motivations or meanings of actors, he "defines the action arbitrarily:"

> If the goal is given, the means follows, and each means then becomes an intermediate goal which must be accomplished by still other means. The total act thus divides into component acts, and an external observer who is "objectively" watching such a series of "component" acts is in no position to say whether the goal has yet

> been reached or whether there is more to come. Each component stage can be regarded as a new unity. (1967:62)

Following Schutz's logic just a bit further, it seems that it would be just as arbitrary for a sociologist to dissect and identify any alleged "component stages" of an "observed action."

Sociologists cannot analyze raw activity, the natural flow of behavior, until its constituent elements have been meaningfully associated or connected by an observer. In a manner similar to Bruno Latour (2005), we argue that sociologists should "trace the associations" observed by an observer. In contrast to Latour, we expect sociologists to clearly identify a particular social system as the observer that assembles associations. John Dewey, the pragmatist philosopher whose influence on symbolic interactionism was profound, described the power of communication to transform and associate raw activity and experience into meaningful elements. "Of all affairs," Dewey wrote, "communication is the most wonderful:"

> That things should be able to pass from the plane of external pushing and pulling to that of revealing themselves to man, and thereby to themselves; and that the fruit of communication should be participation, sharing, is a wonder by the side of which transubstantiation pales. When communication occurs, all natural events are subject to reconsideration and revision; they are readapted to meet the requirements of conversation, whether it be public discourse or that preliminary discourse termed thinking. Events turn into objects, things with meaning. They may be referred to when they do not exist, and thus be operative among things distant in space and time, through vicarious presence in a new medium. (1925:138)

Dewey's fascination with communication stems from his belief that "knowledge is a function of association and communication" and that individual "faculties of observation, reflection, and desire are habits acquired under the influence of culture and institutions of society" (Dewey 1927:158). As we will explain, observers of social systems, both native members of groups and visiting sociologists, inform themselves with associations established by communication, an observer of its own kind.

The Concept of Communication

Social system theory takes an innovative approach to understanding communication. The approach is theoretically sophisticated, abstract, and difficult to simplify. For the sake of clarity, we will introduce this unique concept of communication with the help of a heuristic "bridge" that we will eventually dismantle. The bridge must only be temporarily erected because it is built upon an incorrect premise. We will soon describe communication as the synthesis of information, utterance, and understanding. In the meantime, we will introduce each of these components as if each had a "life" of its own, following an established model of communication that describes a process of "information transfer" between sender and receiver (Shannon and Weaver 1949). This is our bridge. The "faulty" premise that we will eventually reject is that information can be identified independently of the process of communication. As Luhmann noted, "There is no information outside of communication" (2002:160).

Every unit of communication consists of three loosely coupled components: information, utterance, and understanding. The decisive point is that each of these elements must be conceived as a separate *selection*. By calling each component of communication a "selection," we mean that it takes an observer to select where he will direct his consciousness and how he will interpret and react to things he observes in the world. As Luhmann put it,

> Communication grasps *something* out of the actual referential horizon that it itself constitutes and leaves *other things* aside. Communication is the processing of selection… The selection that is actualized in communication constitutes its own horizon; communication constitutes what it chooses, by virtue of that choice, as a selection, namely, as information. What is uttered is not only selected, but also already a selection—that is why it is uttered… Selectivity as such attracts further communication: it recruits communications that direct themselves to aspects that selectivity has excluded. (1995a:140)

Selections of information, utterance, and understanding do not causally determine each other, and each could have been selected otherwise.

We will now describe each of the three components, hoping that readers will hold on to the idea that, when it comes to communication, each one always informs the other two and cannot make an informative difference by itself.

First, the selection of information is always accomplished as a two-step-process that appears to the informed system (the observer) as a single event. Information is formed as "a difference that makes a difference" (Bateson 1972:315). A traffic light, for example, might show itself as red, yellow, or green, presenting a perceivable difference on the level of data. Complementarily, a driver is ready to stop or to go: a difference in possible actions. The red light (difference 1) selects the action "stop" (difference 2). The green light (difference 1) selects the action "go" (difference 2). Or, to say it the other way around, the driver offers the world a distinction that he can control (going or stopping), but he lets the world (here, the traffic light) select one of the two sides, informing the choice that only he can make. A well-socialized driver would never expect the light to change according to his own actions of driving or stopping. Drivers are agents and are free to ignore the information offered by the light. However, to keep a valid license, a driver must permit the light to condition and discipline agency. A driver who refuses to let the light steer the selection of his own possibilities is a driver who may soon be without a license.

The three colors of a traffic light produce a relatively simple set of significant differences that can be selectively connected with a correspondingly simple set of possible actions. The complexity of information, however, can be highly variable and represent a much greater number of possibilities. The more possibilities there are, the more will have to be excluded by selection. The more possible exclusions, the higher the information value of any given data. "By information," noted Luhmann, with reference to the basic terms of social system theory, "we mean an event that selects system states" (1995a:67). In simplistic terms, a "sender" must first select, from all possibilities, what information she wants to send a receiver. She cannot send everything at once. In the next step, she decides how to send out the information to a receiver.

Communication assumes a degree of freedom to choose if the information is shared at all, and if so, in which form. This means that the utterance appears as a second selection. If it were not a selection, communication could not distinguish itself from mere behavior, from mere reactions to external stimuli. In its original sense, the word utterance meant something that had been brought outward. As a component of communication, an utterance may take any form (sound, writing, gesture, significant symbol, etc.) that can bring information (difference) out of a system. A potential participant in communication must select whether or not he will issue an utterance and, if so, what "coded" form it will take. Coded sound, for instance, may appear as spoken language, as words extracted from noise. A different code permits text to appear as writing, as opposed to meaningless black marks on white paper. Partners in communication need to distinguish between coded and uncoded occurrences. "Coded events," Luhmann asserted, "operate as information in the communication process, uncoded ones as disturbance (noise)" (1995a:142).

Neither the selected information nor the selected utterance can determine the third constitutive element, the understanding. The operation of understanding must connect the actualized utterance with other utterances; it must connect the information with other information that was previously brought out. No utterance and no information can stand alone; the meaning of each is supplied always by its relation to something else. In other words, understanding adds context to every utterance and to every piece of information.

Imagine a family enjoying a picnic under a big tree in the country. The father, the mother, and their children do not just sit on the blanket in silence, chewing their sandwiches and staring into space. As "normal" imaginary people, these family members are in nearly constant communication; verbal and non-verbal exchanges are fired off in rapid succession. From the perspective of social system theory, let us try to observe a very tiny part of the communication process in this make-believe interaction.

We see the Mother hold up her empty cup, look into the Father's eyes, and ask, "Is there any more tea?"

The mother has just verbalized a selected bit of information. What are some of the things she had to do in order to make this step towards communication? As a psychic system, she needed to select which bit of information she was interested in uttering. Of all the things she could have possibly done or said, only one thing was actualized in the moment we observed. Before making an actual selection from countless possibilities (for example: I want more food, I want the children to go away, I want to leave, I hate ants and picnics, etc.), the mother did not know what she wanted to say. Drawing a distinction between saying one thing and not saying another is the prerequisite for saying anything at all. To inform, information must always be selected; it must be given form.

The mother's utterance concerning tea was spoken and not sung, written, telegraphed, or otherwise signaled. To get her selected information outside of her own psychic system, the Mother had to select a form of utterance. She had to select a medium of communication. She coupled her information with speech. Why did she choose speech, when so many other ways were available to send information? As observers, we might also wonder how she selected which words to use, which tone of voice, and with whom she should make eye contact. As second-order observers, let us keep the enormous number of possible communication media in mind. If information is to be externalized, one of many possible forms of utterance must be actualized. The mother made her two necessary and contingent choices, first selecting information and then a form of utterance. Completion of the communication now depends on other psychic systems, perhaps on the father or one of the children.

Will anyone present at the hypothetical picnic understand the mother's utterance and offer a selected form of how it might be understood? What are the conditions under which particular forms of understanding might occur? One important general condition is that the recipient must perceive the medium of communication. Suppose the father notices the mother's empty cup and inquiring eyes. Even if he were deaf, he might still be able to read her lips. He returns her gaze and silently nods in affirmation—before quickly falling asleep. The mother sees the nod and draws an inference about its meaning: "Yes, there is more tea." However, now that the father has fallen asleep, the mother may have to pour it herself. She looks at her adolescent daughter, who, still irritated that she was forced to come along on a boring family picnic, gives her a dirty look and turns up the music on her headphones. The younger son sees the mother reach for the teapot, and says, "Can I have the last cup?"

Communication may lead to understanding, but not necessarily to another cup of tea. We can neatly couple our thoughts with information and an utterance, but we cannot guarantee that they will achieve the understanding we hoped listeners would select.

To be understood, a message must be "transmitted" between sender and receiver via an utterance. A sender typically selects a form of utterance that appears likely to achieve successful delivery. As an illustration, suppose a man wishes to send the following message to an unfamiliar woman who attracts his romantic interest: "I am interested in getting to know you." The man may flash a toothy smile, wink in her direction, and repeatedly point to himself and back to her. With a more sophisticated approach, he may send her an eloquent love letter. He might also use his toes to tap out a proposition in Morse code. He might even hire a barbershop quartet to visit the woman and deliver a passionate "singing telegram" on his behalf. Self-referentially considering his alternatives, our man will rationally choose to use whatever form of utterance he thinks will most likely lead to the successful reception of his message.

The time has come to gently take down the aforementioned bridge and give our concept of communication its self-sustaining quality. Our introduction and our examples relied on the implicit assumption that "there *is* information" or "there *is* a message." Luhmann rejected this metaphor of message transmission, asserting "it is unusable because it implies too much ontology:"

> It suggests that the sender gives up something that the receiver then acquires. This is already incorrect because the sender doesn't give up anything in the sense of losing it. The entire metaphor of possessing, having, giving, and receiving, the entire "thing metaphoric" is unsuitable for understanding communication. (1995a:139)

Making the same crucial point, Ernst von Glasersfeld emphasized the difference between the meaning of an utterance and the communication medium that selectively stimulates the perceptions of participants in society:

> The physical signals that travel from one communicator to another—for instance the sounds of speech or the visual patterns of print or writing in linguistic interactions—do not actually carry or contain what we think of as meaning. Instead, they should be considered as instructions to select particular meanings from a list that, together with the list of agreed signals, constitutes the "code" of the particular communication system. From this it follows that if the two lists and the conventional associations that link the items in them are not available to a receiver before the linguistic interaction takes place, the signals will be meaningless for that receiver. (2008:62)

Referring back to our illustration of a traffic light, the light offers "a list of agreed signals," each of which is symbolically associated with a second list of possible actions (albeit only in the prepared minds of cultured observers). The action of stopping, of course, does not exist *in* the red light. The message does not exist in itself: it must be imagined.

When granted its own ontology, as in the traditional model of communication, a message appears to be a thing that possesses distinct margins and obvious import. Additionally, a "guardian" is assumed to be present, controlling this thing, this "message," carefully watching to make sure that it is transferred accurately, guarding against "information loss," and ensuring that it retains the distinct qualities it had at the beginning of the communication process. The "guardian" might be first the sender, then the receiver. As long as the process of information transfer is guarded from interference and pollution, understanding is typically expected. However, as Luhmann maintained, "Understanding normally includes more or less extensive misunderstandings; but these are always, as we shall see, misunderstandings that can be controlled and corrected" (1995a:141).

To view understanding as a constitutive element of communication, and not as a secondary effect of communication, is already an important shift from the traditional conception. But the most innovative and difficult modification lies in the conception of communication as a synthesis of three elements. Communication is not "information + utterance + understanding." Instead, communication is the synthesis of all three selections into a single unit. Whenever communication occurs, all three elements have been successfully combined—without the attention of a guardian. If communication fails, none of the elements gains any social or communicative significance.

Communication takes the form of a synthesis because no individual could surprise himself with his own information, persuading himself with an appealing utterance and surprising himself with a new understanding—all realized in one single event. The impossibility for every single being or any other force to accomplish this synthesis is the reason for Luhmann's often attacked, but seldom understood, assertion: "Humans cannot communicate... Only communication can communicate" (2002:169).

Connecting Utterances with Meaning

How, in the absence of any external producer and controller, can we theoretically describe this synthesizing form of communication? The solution is to conceive the utterance as a constitutively divided, "partialized," or split event. One part of the utterance introduces the information: any news about the world or about the system's state. The second part simultaneously enacts a test of the understandings supplied in previous utterances. The third part anticipates the occurrence of future utterances (and understandings). Thus, it is up to the utterance to represent the mysterious trinity of communication as one single event. We see or hear only utterance after utterance after utterance—but we can distinguish them from noise only by introjecting information and understanding, by imagining the self-reference of psychic and social systems. The utterance can then be grasped as the self-simplification of communication. Every utterance reduces communication to an event that can be easily perceived, but communication itself cannot be directly observed—it can only be inferred from utterances (Luhmann 1995a:164). The complexity of communication is maintained with the question: which utterance will be the next one? By conceiving of an utterance as a representative of communication, we relieve ourselves of the problem of having to read the fleeting intentions, goals, and interests of actors. Instead we can focus on the connectivity of communication itself.

We confess that this account of communication may be hard to understand. The familiar model of information transfer is much easier to grasp. However, the theoretical sophistication of this unusual "three-in-one" concept of communication opens up fascinating new possibilities for analyzing social processes. As an initial reward for taking the difficult step towards this alternative perspective on communication, social system theory offers a much more interesting problem than that of how individuals accurately transfer information. This new view on communication invites the question of how one can ever become eligible or viable as sender or receiver. Who has a right to speak to us? Who can be ignored? Who has the right to receive confidential information? When that person talks, why do people listen? According to Luhmann,

> To communicate, humans must become persons, they must become an address for communication. Attaining and keeping an address within a particular communication event requires each participant to accept the contingent forms of communication constructed by the particular social system in use. (2002:182)

The problem of establishing oneself as an address for communication begins very early. An infant cries without controlling any distinct margins around its intention or information. It is decisive that *somebody* discriminates its cry from mere noise and begins to guess and bet effectively what the information might be. Guesses and bets are confirmed by a baby's smile—or rejected by the next cry.

Managers and Assessors: An Ethological Model of Communication

To illustrate the problem of selecting addresses of communication, we can refer to an ethological model of communication. Donald Owings and Eugene Morton described a new conceptual framework for understanding vocal communication among animals that supplants the traditional information transfer model discussed above (1998; see also Hauser 1996). According to Owings and Morton, communicative systems emerge from "the dynamic interplay" between two equally active roles, that of manager and assessor:

> All individuals play both types of roles in communication, and the logic of natural selection indicates that individuals should be selected for their effectiveness in both. As managers, they should be able to use communication to achieve fitness-enhancing ends by influencing the behavior of others, in part by exploiting their assessment systems. As assessors, they should be capable of making adaptive behavioral adjustments through selective attention to the most reliable cues available appraising individuals and situations, whether or not these cues arise from signals. (1998:29)

In the assessment/management model, communication is *not* conceived as the exchange of information between sending and receiving animals. Rather, communication is one very effective means of assessing and managing the behavior of others "via signaling so as to maintain or attain a preferred state;" it is a means that can substitute for dangerous fighting over control of resources (Owings and Morton 1998:235). According to this ethological model, communication "focuses equally on assessing and managing, seeking insights into the success of managing by exploring the properties of assessment" (1998:242–243). While Owings and Morton focused on animal communication, it should be noted in passing that the linguist Arie Verhagen recently formulated an account of communication that also stresses the pragmatic goals of assessment and management. Verhagen wrote, for instance, that the "default condition for ordinary expressions is that they provide an argument for some conclusion, and this argumentative orientation is what is constant in the function of the expression, while its information value is more variable" (2005:10).

Illustrating their model, Owings and Morton described the way in which female hooded warblers assess the communication of male managers. Male birds compete for both food-rich territory and mating partners.

These are two of the most important regulatory problems that individual males seek to manage. According to the logic of natural selection, performance as managers defines the relative fitness of males for survival. Female birds evaluate both a male and the quality of his territory by assessing the quantity of time he spends singing. Strong males have ample time to sing because they command territory with plenty of available food. While female assessors pressure males to sing, males search for territory that permits them to sing as much as they can—without ever attempting to transmit information. The message is the song, we might say. Once a female selects a male and his territory, the male stops singing his unique song and turns to manage a different regulatory problem: keeping other males away from his mate. This is not easy because the female partner seeks to manage her own chief regulatory problem, the certain fertilization of her eggs. As she builds a nest, lays eggs, and incubates those eggs, she chases other females out of the territory and attracts other males by producing a "chinking" sound. It is typical for a female bird to manage a male into feeding fledglings sired by a different male (Owings and Morton 1998:4–5).

As sociologists we need to only adjust the description and the interpretation of the example of communication among birds. First, let us replace "the logic of natural selection" by "the logic of communicative selection." Communication, then, is the selection of effective managers and assessors. We can also say that communication is the reproduction of the difference between managing and assessing, because neither side can be identified without the other. The result of the selection is not survival in a biological sense, pertaining to the survival of organisms and species. Rather, the result of the selection is in being or not being a person; having an address; making differences for and within communication. Regarding their regulatory problems, male and female birds might struggle for individual control—like humans seeking to control their interests. But communication registers merely the difference between singing and not singing, reproducing the ongoing coupling of managing and assessing. Communication identifies managers and assessors; that is, persons who make a difference for and within certain social systems. No individual can ever function as assessor or manager in every social system. No social system could ever select all individuals as addresses. Total inclusion would sacrifice any form of boundary and, therefore, any internal connectivity.

In a manner analogous to that of social system theory, the assessment/management model rejects the idea that information exists, and that it is transmitted from sender to receiver through communication. Participants in communication do whatever they do—singing, chinking, barking, etc.—because they are interested in controlling their environment, regulating their own affairs, and remaining viable operators. Successful methods of management and assessment are repeated by individual animals that survive to pass on their genes. Communication selects behavior that reproduces the possibility of further communication, thereby making repetition probable. Both theoretical approaches emphasize the importance of documenting and understanding the structural conditions under which communication becomes possible. Rejecting any ontology of information, the two approaches share a primary interest in the ontogeny and evolution of the physiological or technical structures required for making and perceiving signals and signs. In other words, both the ethological and sociological theory seek to understand the available forms of communication, or the structural means for indicating and perceiving differences.

There are other important similarities between the assessment/management paradigm and social system theory. For instance, both theories begin by assuming that the evolution of a species or system has no particular teleology, goal, or final destination (Luhmann 2002:172; Owings and Morton 1998:57). Both theories emphasize that a species or system builds and self-referentially organizes its own environment, constrained by the limits of its own perceptual and noise-producing abilities (Owings and Morton 1998:60). As we stated above, social system theory explains that understanding the indication of a communicative utterance involves recursive indexing and cultural semantics. Referring back to the illustration of the assessing driver and the managing traffic light, we can indicate a cultured coordination of two forms of differences. According to Owings and Morton, the ethological theory asserts that the "proper place of semantic information is clarified through reference to the sharp distinction between management and assessment:"

> Information is useful shorthand to describe what assessment produces. In strong contrast, management is focused on producing behavioral changes, not information… When the active nature of assessment was not recognized explicitly, the transfer of information was needed to account for the impact of signals on the behavior of others. Information, sent by signalers and received by targets, took on an active, causal role. But, since individuals are already active in assessment, then nothing needs to be "sent" by managers to make assessing individuals behave: *they are already behaving in their own interests.* Therefore, managing individuals try to influence *how* assessing individuals behave by affecting the conclusions they reach in assessment. Management is focused on pragmatic outcome, not information, both proximately and ultimately. (1998:54–55)

Social system theory and Assessment/Management theory are both concerned with describing the structural and semantic conditions under which communication occurs. When communication does occur, both theories pursue the pragmatic questions: how was it observed and what did it accomplish?

THE SELECTIVITY OF COMMUNICATION

It is impossible to predict what communication can accomplish. One might imagine society having a selfish interest in its own survival and growth, reproducing itself by provoking, manipulating, and managing the motivations and needs of psychic systems. As it helps participants accomplish their goals, society achieves what constructivists might recognize as society's own viability. In the course of reproducing itself in the continuous course of communication, society processes meaning independently of the moral sensitivities, normative concerns, and group loyalties that tend to interest its participants. For its part, sociology has traditionally taken a strong interest in marginalized social groups, in "outsiders" and criminals. Deviants are frequently described as if they had slipped outside of society, as if they occupied an excluded place. And society is traditionally identified with its most morally correct, economically prosperous, politically privileged, fully included participants. Social system theory, however, asserts that society equally excludes all people, while at the same time helping all kinds of people, groups, and organizations accomplish their own goals. Society is truly value-free. Good guys and bad guys, cops and robbers, saints and sinners, rich and poor: as equals, all turn to communication for assistance in negotiating their needs. Every understanding adds a new link to the chain of communication.

Terrorists, spies, thieves, liars, revolutionaries, and impostors are all able to order pizza, attend a football game, attend a church picnic, and participate in all kinds of other normal social activities. The notorious hijackers who took control of four American passenger airplanes on September 11, 2001, would not have succeeded if security agents had stopped them from boarding the planes. The terrorists, of course, did not communicate their intentions. Prior to their attack, they had been living normal lives in the United States; going to restaurants, dating girlfriends, buying designer clothes, drinking at bars, and making financial arrangements at banks. The terrorists routinely participated in communication with Americans—just not about being terrorists. They were insiders of American communication, even while thinking of themselves as outsiders of American culture. Selectively taking part in society, the terrorists accomplished many different things with those they eventually destroyed. To consider how a group of dead terrorists embraced society in order to achieve their martyrdom, their own construction of viability, underscores the morally blind nature of sociology's subject.

Just as the patriot and the traitor can join each other in saluting the flag and accomplishing the performance of loyalty, people from different countries or territories can share in society. In this era of globalization and worldwide associations, it is normal for communication to accomplish the creation of society across spatial, political, economic, and regional borders. According to Luhmann, sociologists have found it difficult to stop thinking about societies as political or territorial units, differentiated clearly within the dimension of space. Nonetheless, it is no longer plausible to think of Brazil as constituting a completely different society than Thailand, the United States as a different society than Russia, Uruguay as a different society than Paraguay (Luhmann 1997a:25). Extending Luhmann's observation, we know that American dollars effectively accomplish the satisfaction of desires in each of these countries. Men and women use different public bathrooms here and there. AT&T can make the necessary telephone connections. People drink Coke. Drivers for UPS drive the same brown vans and wear the same brown uniforms. A man in Milwaukee buys a tuba on eBay from a lady in Taiwan. South Americans convert to Mormonism and look toward a sacred Salt Lake City, hearing the tabernacle choir perform live via satellite. Communication draws its own map of accomplishments.

Spatial boundaries no longer limit what communication can accomplish, and, therefore, can no longer be thought to contain or represent societies. This does not imply that a Russian and a Paraguayan will be able to easily understand each other. For that matter, it does not imply that two Russians will understand each other with inherent ease. Regardless of how much people believe they have in common, successful communication is always improbable. There are just too many things that can go wrong. If individuals do manage to successfully participate in communication, it is *not* due to normative consensus or common regional, national, or spatial origins. Rather, it is by employing established systems of communication to successfully manage other participants.

Thoughts alone can hold onto thoughts, and only communication can hold onto communication. Within social system theory, this condition is known as *autopoiesis,* when systems reproduce themselves with their own elements. Explaining this concept, Luhmann cited the biologist Humberto Maturana:

> An autopoietic system is a system with a changing structure that follows a course of change that is continually being selected through its interaction in the medium in which it realized its

> autopoiesis… An autopoietic system is either in continuous structural coupling with its medium or it disintegrates. (Luhmann 1995a:220)

Structurally coupled with the thoughts of humans (and with life, sound, light, and other possible media), communication accomplishes understanding in a recursive, autopoietic manner (compare with Maturana 2002).

Psychic and social systems orient themselves with the past, referencing memory to make sense of their own histories in the present. "Memory," in Hume's opinion, "is but the faculty by which we raise up images of past perceptions" (1978:261). What memory does with thoughts, social systems do with culture or social memory.[1] According to their own self-referential meaning-making system, both types of systems attempt to hold on to selected thoughts or communications by recursively linking them with other events from the past. As Peter Fuchs maintained,

> At any moment, the events that have passed acquire a new identity, a new meaning. Considering these circumstances, we can conclude that the three selections we have mentioned (information, medium, and understanding) are bound to time… The most important (but also impenetrable) consequence is that no social system has access to its own actuality. We may also say that they are blind to their own presence. They are reality-constituting systems because they never find their present essence, their intrinsicality, or their substance. They have to declare what has taken place, and they are unable to describe what their reality is. (Fuchs 2002a:4)

Culture and memory crystallize certain forms of communication and thought, preserving them in the same form of operation—as either communication or thought—without substantive or isomorphic reproduction.

Becoming and remaining a viable observer depends on learning behaviors that work and discovering strategies that successfully manage forces in the environment. Much of what is learned comes through one's personal experience of success and failure. Consulting social memory, however, yields a far greater amount of knowledge than any one person could amass.

Kingsley Davis asserted,

> Not only does learning become more rapid under a system of symbolic communication, but the total volume of things learned is increased many thousand times. If we take into account the techniques, rules, rituals, beliefs, names, vocabulary, attitudes, anecdotes, statuses, and facts that each person learns, we see that the total amount even for the most primitive society is infinitely beyond what an anthropoid—or, for that matter, a cultureless human being—could acquire. (1949:45)

Acquiring culture and memory through participation in communication speeds up the process of gathering and sorting out information about one's environment. Luhmann suggests that the evolution of human beings can be characterized by two simultaneous developments: "extreme social dependence and marked individuality" (1997a:193). Individual systems of consciousness do not come to directly depend on one another; rather, they come to depend on a higher order system. "For human beings," Luhmann noted (1997a:194), "this higher order system, that does not itself live, is the communication system of society:"

> Participants can bring into communication their own perceptions and the interpretations of the situation that are bound up with them, but only according to the autonomous laws of the system of communication, for instance, only in the form of language, only by claiming speaking time, only through imposing oneself, making oneself visible, exposing oneself—thus only under discouragingly difficult conditions. (Luhmann 2002:158)

Even with the odds stacked against its success, communication is the only way that cultured psychic systems can reach an understanding. There can be no social action, no society, until communication establishes associations and connectivity. Thus, communication, not actors or interactions, is the fundamental unit of sociological analysis.

Endnote

1. See Luhmann's treatment of time and memory (1997a:576–594) and our discussion of culture in the previous chapter.

Chapter 3

Psychic Systems

Subjectivity, Objectivity, and Understanding

Peter Gay, the interpreter of Freud and psychohistorian who taught at Columbia and Yale, once wrote an article about thinking and feeling in the modern world. His piece began with an extraordinary question:

> When did our minds become "our" minds? Our fears of overpopulation, of race riots, of wasted resources, of extinction through nuclear war, are recent in origin, at least in their acute form. But the minds that furnished our minds reach farther back: practically all the ideas that determine our vocabulary, direct our taste, dominate our lives, were generated or revived in the years between 1890 and 1914. (1973:10)

Gay goes on to eloquently describe the powerful philosophical, scientific, and artistic currents that marked the brief period of time before the First World War. As marvelous as it may have been, the Zeitgeist of that period was not as extraordinary as Gay's assumption that "our minds" have anything at all in common. Indeed, how could one mind furnish another mind with anything? Common sense seems to suggest that each person's consciousness is a completely confidential affair and that each subject thinks with his or her own mind on his or her own terms.

From his vantage point as a philosopher of mind, John Searle argued "consciousness is a biological phenomenon," "part of our ordinary biological history, along with digestion, growth, mitosis, and meiosis." However, as Searle asserted, consciousness has an important feature that other biological phenomena do not have: "subjectivity." He continued: "There is a sense in which each person's consciousness is private to that person, a sense in which he is related to his pains, tickles, itches, thoughts, and feelings in a way that is quite unlike the way that others are related to those pains, tickles, itches, thoughts, and feelings" (1993:4).

Social system theory takes the position that each person represents a private black box of subjectivity, inside of which a meaningful world is independently constructed in a cognitive process that is hidden from others. An actor may observe another's behavior, but not the mental processing that motivates, executes, and connects that behavior with subjective meaning. The veiled consciousness of independent subjects should logically lead away from any orderly association of minds, shared meanings, or mutual understanding. Nevertheless, as this chapter demonstrates, scholars in various disciplines have repeatedly turned to notions of "intersubjectivity," "group mind," or "collective conscious" to explain how isolated subjects are mentally and behaviorally coordinated. Sometimes intersubjectivity is used in a merely figurative sense, with a claim that individuals understand one another *as if* they possessed a single, mutually accessible mind. In other instances, intersubjectivity is used in a literal manner, with the assertion that individuals may actually read or observe one another's thoughts. In this chapter, we distinguish between various ways the concept has been used, and evaluate the potential of intersubjectivity to inform and order the thoughts of individuals.

The concept of intersubjectivity generally implies that a common form of consciousness exists between or among different actors and that "our mind" provides us with a mutually or reciprocally accessible source of meaningful information. Conditioning our private thoughts with intersubjectivity, as it were, we simultaneously think with and for one another, jointly informing ourselves with the same cognitive resource. It makes sense, along these lines, to inquire about the collective psychology of a historical epoch. Intersubjectivity is a familiar notion within the disciplines of sociology and psychology, two fields that share a central interest in describing how individuals make sense of one another. More recently, the discovery of "mind reading neurons" has incited intense research activity among neuroscientists. After considering some of the most influential positions on intersubjectivity, we conclude that the concept should be rejected. Those who propose it fail to take the subjectivity or the intentionality of individual actors seriously, they confuse perceiving another person's behavior with understanding its meaning, or they propose that actors have clairvoyant powers of extra-sensory perception that are not documented. We suggest that sociology's traditional attraction to the notion of intersubjectivity results from its struggle to develop a scientific concept of society that adequately accounts for the social organization of autonomously thinking subjects. Building on two enlightening insights from social system theory, namely that "society is communication" and "society excludes subjectivity" (Luhmann 1997a), we explain how communication coordinates the independent minds of its participants. The operative closure of conscious systems does not lead to social chaos; this is because of the unlikely coordination produced by society. Subjects teach themselves to condition their private thoughts, intentions, tastes, and behaviors with interobjective, self-referential forms of meaning that are established in communication. The subjectivity of each participant in society remains sovereign, opaque, and closed.

Subjectivity as a Problem

In the familiar humanistic tradition, subjectivity is attributed to each individual person and is typically described as the mental capacity for reflexive thought, self-awareness, or self-consciousness. In *Phenomenology of the Spirit,* for example, G. W. F. Hegel described the subject's longing for and alienation from objects in the external world (1977). His "biography of the spirit" is a story about the subject's eventual reconciliation with the "discipline of reason." To grossly simplify Hegel's conclusion, the reasonable subject is the one who resolves all contradictions between personal and communal goals, conditioning subjectivity with culture, morality, religion, science, and the state. The image of individual people working in harmony to make a common home in the world is appealing, but what power can guarantee that different subjects will experience the same spiritual biography? Hegel's readers seem to have struggled between two possibilities. On the one hand, the state might organize and enforce social order, negatively sanctioning any unreasonable subjects. On the other hand, the social process of mutual recognition and reciprocal self-determination might lead individuals to voluntarily fall into spiritual alignment. Perhaps all will simply recognize the truth that one's own freedom depends on the freedom of others, and that one must internally want to do only what others expect one to do. One way or another, the problem of organizing independently thinking subjects is covered up by Hegel's assumption of a common human consciousness, a subjectivity of the reasonable herd that grazes through history. Subjects either lose themselves in the external identity of the state, or they are internally guided by the same state of absolute truth.

The sociological pioneers seem to have followed closely in Hegel's footsteps. Karl Marx, for example, claimed the ability to recognize the difference between true and false consciousness. Every truly conscious subject identifies with the working class. Marx knew the needs of all workers, enlightened as he was, so he also knew the needs of every truly human subject. The dictatorship of the proletariat will be replaced by a common internal discipline, a shared realization that class-consciousness is the only authentic consciousness for the "species-being." As "reformers of consciousness," Marxists "only show the world what it is fighting for, and consciousness is something that the world *must* acquire, like it or not" (Marx 1978:15). As truly conscious subjects, workers will know that they want to consume only what they absolutely need, while contributing to the community as much as they possibly can. The smooth organization of communist society is assured, it seems, by laws of nature. There is simply no stopping history's rush toward peace and harmony. Juergen Habermas (1984), sustaining this familiar vision, asserts that individual subjects can talk their way into agreement with an appeal to "communicative rationality." Whenever participants in communication mutually acknowledge one another's human dignity and right to freely express subjective opinions, the "ideal speech situation" leads the group to consensus. When subjects are respectful to each other and talk long enough, reason descends upon them and social order prevails. If the world cannot stop itself from acquiring true consciousness and if communicative rationality cannot fail to achieve understanding, why worry about convincing subjects to join the ideal speech revolution? The consciousness reformers claim that those who cling to their own subjectivity lack revolutionary zeal and communal spirit. They lament that so many individuals are still guilty of falsely thinking for themselves, irrationally resisting the laws of nature and the forward march of history. However, if subjectivity is taken seriously, would not the difference between true and false consciousness be a purely psychological, aesthetic, or introspective concern?

Emile Durkheim took the problem of maintaining social order seriously, as is evident in his *Rules of Sociological Method* (1938) and other works. Individuals will do whatever they want unless they are constrained by a disciplining moral community. In a quasi-behaviorist fashion, "social facts" condition individual behavior because members of the community expect that deviants will be publicly punished and conformists will be rewarded. Social facts determine the actions, thoughts, and even feelings of individual subjects, as Durkheim maintained:

> [Social facts belong to] a category of facts with very distinctive characteristics: it consists of ways of acting, thinking, and feeling, external to the individual, and endowed with a power of coercion, by reason of which they control him. These ways of thinking could not be confused with biological phenomena, since they consist of representations and of actions; nor with psychological phenomena, which exist only in the individual consciousness and through it.

> They constitute, thus, a new variety of phenomena; and it is to them exclusively that the term "social" ought to be applied. (1938:60)

When behavioral norms are no longer enforced by sanctions, a pathological condition of anomie results and members lose the ability to predict one another's behaviors. Durkheim recognized that subjectivity exists, that it implies independence for members of society, and that this independence presents a challenge to social solidarity and order. In his theoretical account of social integration (1964), "mechanical solidarity" works as a source of order in primitive society because the individuality or subjectivity of each member is very weak and social differentiation is very low. The common denominator appears to be provincialism, for each member of the primitive community knows only as much as every other member (see Redfield 1953). For its part, "organic solidarity" works in modern society because subjectivity is restrained and informed by membership within differentiated, specialized social networks. With both forms of Durkheim's social solidarity, subjectivity is shaped by a "moral community" that is inspired by a "collective consciousness." A common awareness of sanctions is what appears to establish the social facts that constrain the behaviors, thoughts, and even feelings of actors, but Durkheim does not explain how the collective consciousness informs each actor's consciousness. Franklin Henry Giddings, one of the first sociologists in the United States, integrated Durkheim's reasoning into his own conception of "the social mind." He asserted:

> The social mind is a concrete thing. It is more than any individual mind and dominates every individual will. Yet it exists only in individual minds, and we have no knowledge of any consciousness but that of individuals… The social mind is the phenomenon of many individual minds in interaction, so playing upon one another that they simultaneously feel the same sensation or emotion, arrive at one judgment and perhaps act in concert. It is, in short, the mental unity of many individuals or of a crowd. (1896:134)

Giddings observed that the social mind takes deliberate action to "impose its standards, codes, policies, faiths, and creeds upon individuals"

(1896:152). With a Hegelian twist, he claimed that when "the group of associated persons becomes a community," there emerges "a comprehension by each mind of some portion of thought and feeling of all other minds. In this phenomenon lies the possibility of a perfect social coordination without the sacrifice of individual freedom" (1896:390). Again, it is not clear whether individuals conform to the social mind because they commonly recognize its higher morality or reason, or because they feel pressured to join the crowd. Durkheim's description of a community of morals also informed Talcott Parsons theory of normative consensus (Parsons and Shils, et al. 1951). If a community appears to be functioning, we may presume that its members share an underlying sense of morality. But is there not ample evidence that common immorality can function as a form of counter-cultural morality, integrating and organizing individuals (Alexander 2001; Schneider 2001; Kingma 2008; Brents and Hausbeck 2001)? When associated in moral or immoral communities, how do individuals mutually order themselves if each one thinks alone?

Perhaps more than any other pioneering sociologist, Max Weber (1972) directed his attention to the sociological implications of subjectivity. His notion of sociology's project, "to explain and understand social action," explicitly pursues intentionality or the meaning subjects attach to their actions (Weber 1972:1). Actors connect means and goals with subjective meaning; they select behaviors that they expect will lead to selected ends. As a sociologist, Weber sought to understand the causes of behavior in terms of the subjective motivations of actors, as opposed to the laws of nature, universal reason, or history. The motivations themselves appeared more or less predictable, common, and immanent: physical well being, psychological comfort, economic success, political power, scientific insight, etc. Nonetheless, Weber's key insight was that the connection between behaviors and motivations is subjectively determined. On one hand, multiple participants in the same social action may be moved by different motivations. On the other hand, the same "ideal-typical" motivation may lead participants to engage in different types of action. In other words, knowing a subject's motivation does not enable the sociologist to predict behavior, and seeing a subject's actions reveals nothing about internal motivation. For Weber, subjectivity is like a wild card that draws its own values. To explain how subjects ever come to value the same associations between means and ends, he turns to Nietzschean metaphors of struggle, battle, and competition between charismatic prophets and heroes who invent meaningful lifestyles for other people to lead (see Fleischman 1964). The modern subject either decides how to act so that life gains meaning, creating a personal god, or else decides to find meaning in accepting another subject's commands. In any case, it is up to an individual to choose the meaning of life and a corresponding lifestyle. How will Weber's model reach the subjective world of another, how will his model understand it? Apparently, Weber's methodology was to reduce the motivations of other subjects to his own ideal-typical constructions, filtering away any unrefined idiosyncrasies. Alfred Schutz (1967:234–235) drew attention to this problem of Weber's methodology:

> It seems contradictory to set up the sociologist as judge of what is meaning-adequate, unless we mean by [Weber's] "knowledge of average habits of thought and feeling" the knowledge the social sciences have of all conceivable subjective experiences whatsoever. It is enough for the meaningful interpretation of another's behavior that I assume that my ideal construct stands in the context of meaning for *him*.

The possibility of adequately understanding the meaning behind another actor's behavior is not only the central task of interpretive sociology, it is a task set before every participant in every social interaction. The burning question for every actor is, "How can I assume that the same ideal construct frames meaning in both my mind and that of my partner?"

From Marx, Durkheim, and Weber, sociology inherited an increasing awareness that subjects are subjects because they have intentions: subjects make decisions. One mind at a time, they decide to join the revolution and share true consciousness; they decide to accept social facts and are guided by the collective conscious; they decide to choose connections between means and ends and thus give their lives meaning. As human subjects are assigned more and more responsibility for making their own decisions and defining their own situations, the social order appears to be increasingly improbable and difficult to explain. In a variety of forms, the concept of intersubjectivity has been introduced as a possible solution by several influential writers.

Intersubjectivity as the Basis for Understanding

From Max Scheler's perspective, a subject may understand another by *directly perceiving* the other's mental and emotional states:

> For we certainly believe ourselves to be directly acquainted with another person's joy in his laughter, with his sorrow and pain in his tears, with his shame in his blushing, with his entreaty in his outstretched hands, with his love in his look of affection, with his rage in the gnashing of his teeth, with his threats in the clenching of his fist, and with the tenor of his thoughts in the sound of his words. (1954:254)

In their much-acclaimed treatise on the sociology of knowledge, *The Social Construction of Reality*, Peter Berger and Thomas Luckmann reiterated Scheler's position:

> Human expressivity is capable of objectivation, that is, it manifests itself in products of human activity that are available to both producers and to other men as elements of a common world. Such objectivations serve as more or less enduring indices of the subjective processes of their producers… For instance, a subjective attitude of anger is directly expressed in the face-to-face situation by a variety of bodily indices—facial mien, general stance of the body, specific movements of arms and feet, and so on. These indices are continuously available in the face-to-face situation, which is precisely why it affords me the optimal situation for gaining access to another's subjectivity. (1966:34)

Subjects can perceive one another's thoughts and inner states because they recognize expressive "objectivations." The body reveals the mind.

If there was ever any doubt about the meaning behind a display of another's body, we may turn to the "tenor of his thoughts in the sound of his words." The thoughts are in the expression, or so it would seem.

Edmund Husserl's work on intersubjectivity encouraged sociologists such as Alfred Schutz, Peter Berger, and Thomas Luckmann, to emphasize the "Lifeworld's" ordering influence on the consciousness of subjects. Primarily through their use of language, subjects "internalize" a "symbolic universe" of social structures, institutions, and legitimations in the course of organizing everyday life with others. As Berger and Luckmann argued,

> This nomic function of the symbolic universe for individual experience may be described quite simply by saying that it "puts everything in its right place." What is more, whenever one strays from the consciousness of this order (that is, when one finds oneself in the marginal situations of experience), the symbolic universe allows one to "return to reality"—namely, to the reality of everyday life. Since this is, of course, the sphere to which all forms of institutional conduct and roles belong, the symbolic universe provides the ultimate legitimation of the institutional order by bestowing upon it the primacy in the hierarchy of human experience. (1966:98)

From the phenomenological perspective, intersubjectivity is produced when subjects internalize the human created objective social reality of their shared "Lifeworld," inform themselves with a "social stock of knowledge," and guide themselves with the same "public definition of being." The onus on this position is to explain how the process of internalization works: how does the ordering consciousness of the symbolic universe that "puts everything in its right place" get inside multiple subjects? It would be tempting to accept Berger and Luckmann's claim that socialization is the key, but this only describes the same problem with another word. How does society get inside subjects, how do subjects become internally social?

The early symbolic interactionist school developed a notion of intersubjectivity that seems close to mind reading (Scheff 2006). In the case of George Herbert Mead, for instance, intersubjectivity was suggested by the capacity to "take the role of the other." According to Mead,

> [thinking] is nothing but the response of the individual to the attitude of the other in the wide social process in which both are involved, and the directing of one's anticipatory action by these attitudes of the others that one does assume. Since that is what the process of thinking consists in, it cannot simply run by itself. (1934:260)

Mead emphatically rejected the idea that a subject can create its own meaning: "Our symbols are all universal. You cannot say anything that is absolutely particular; anything you say that has any meaning at all is universal" (1934:146). Mead positioned the subject in a state of complete dependence on society:

> The individual enters as such into his own experience only as an object, not as a subject; and he can enter as an object only on the basis of social relations and interactions, only by means of his experiential transactions with other individuals in an organized social environment. [T]he existence of private or "subjective" contents of experience does not alter the fact that self-consciousness involves the individual's becoming an object to himself by taking the attitudes of other individuals toward himself within an organized setting of social relationships, and that unless the individual had thus become an object to himself he would not be self-conscious or have a self at all. (1934:225)

Mead demonstrated an initial awareness of the function communication plays in the production of meaning and understanding. Communication using significant symbols, as he described it, is a process of "putting one's self in the place of the other person's attitude." In a remarkable passage, Mead wrote,

> Remember that what is essential to a significant symbol is that the gesture which affects others should affect the individual himself in the same way… If that system of communication could be made theoretically perfect, the individual would affect himself as he

> affects others in every way. That would be the ideal of communication, an ideal attained in logical discourse wherever it is understood. The meaning of that which is said is here the same to one as it is to everybody else. Universal discourse is then the formal ideal of communication. If communication can be carried through and made perfect, then there would exist the kind of democracy to which we have referred, in which each individual would carry just the response in himself that he knows he calls out in the community. That is what makes communication in the significant sense the organizing process in the community. (1934:327)

In the perfect society, it appears that subjects would communicate without experiencing information loss, deficits in logic, or unexpected misunderstandings. The commonly available significant symbols would entirely satisfy the particular concerns of every subject.

In Charles Horton Cooley's formulation, each subject lives "in the minds of others without knowing it, just as we daily walk the solid ground without thinking how it holds us up" (1922:208). In his view, "all mind hangs together in a vital whole, from which the individual is never really separate;" "every thought we have is linked with the thought of our ancestors and associates, and through them with that of society at large" (1907:97). Cooley suggested that one individual might understand another through "sympathetic introspection" or by adopting the other's perspective of the world. He directly grounded understanding in communication, by asserting that sympathy "denotes the sharing of any mental state that can be communicated" (1922:136). He described communication as "a system of standard symbols existing for the mere purpose of conveying thought." If one does not participate in communication, he stated, "the mind does not develop a true human nature, but remains in an abnormal and nondescript state neither human nor properly brutal" (1909:62). One's "mind is a microcosm of the society" to which one belongs (1922:144). For Cooley, "society is an interweaving and interworking of mental selves. I imagine your mind, and especially what your mind thinks about my mind, and what your mind thinks about what my mind thinks about your mind. I dress my mind before yours and expect that you will dress yours before mine. Whoever cannot or will not perform these feats is not properly in the game" (1927:200–201). In Cooley's writing, one may recognize a precursor to Erving Goffman's account of the "presentation of the self in everyday life," a self that actively and strategically manages the impressions it seeks to create in the minds of others (1959). Goffman's dramaturgical approach to the presentation of the self in everyday life seems to extend Cooley's metaphor of the "looking-glass self." In Cooley's words, "As we see our face, figure, and dress in the glass, and are interested in them because they are ours, and pleased or otherwise with them according as they do or do not answer to what we should like them to be, so in imagination we perceive in another's mind some thought of our appearance, manners, aims, deeds, character, friends, and so on, and are variously affected by it" (1922:184).

Mead, Cooley, and Goffman shared a sense that actors expect and look for feedback on their performance from an audience of other subjects. Describing how actors do this, they used metaphors that imply mind reading; but they did not assert that subjects could actually access one another's thoughts. If taking the role of the other is a private guessing game that yields only a more or less approximate model of another's mind, exactly how does information make its appearance in the "looking-glass?" The realization that intersubjectivity was impossible led symbolic interactionists to appreciate the importance of the mutually accessible symbols one can see others also observe, such as those used in communication. Even without access to the thoughts of one's audience members, one might detect whether they were clapping, booing, or throwing tomatoes. Despite the attention they devoted to the organizing effect of communication, symbolic interactionists were unable to identify and differentiate its unity as society. With their creative use of concepts, such as social consciousness and social mind, they confounded the difference between private and public meaning. The words they used to describe communication indicated subjective qualities of mind and consciousness, diverting sociology's attention away from the objective resources of communication and towards the inner experience of the individual.

With perhaps a more desperate need than sociologists can experience in their work, mental health professionals must cope with the "subjectivity gap" between themselves and clients when providing therapy. Sigmund Freud, of course, understood that interpreting the dreams of others is difficult because there is no fixed or common vocabulary of symbols in use. The same content might mean different things to different dreamers (Freud 1987:96). From another vantage point, Carl Jung (1959) suggested that the same fixed primordial images or "archetypes" appear in dreams and fantasies because individuals have inherited a common human psyche, a "collective unconscious" or "objective psyche." "If we had to interpret dreams in an exhaustive way according to scientific principles," Jung contended, "we would have to refer every such symbol to an archetype." In a therapeutic context, in careful consideration of the patient's psychological state, the meaning of symbols in a dream should be treated "as if they were not fixed" (Jung 1955:178). It seems that the consequences of the subjectivity gap are exacerbated by the therapist knowing more about the patient's subjectivity than the patient.

Their differences aside, Freud and Jung agreed that psychoanalysis could not proceed unless clients and therapists were willing to engage in communication. Building on this insight, contemporary psychoanalysts recently have proposed a new form of therapy rooted in an "intersubjective perspective" (Bradfield and Knight 2008; Stolorow 1994, 2000; Stolorow and Atwood 1992). According to Robert D. Stolorow, a psychiatrist at UCLA's School of Medicine,

> Clinical phenomena, such as psychopathological states, transferences, resistances, and negative therapeutic reactions are grasped, not as products of intrapsychic mechanisms originating within the interior of the patient's isolated mind, but as taking form at the interface of the interacting experiential worlds of patient and therapist. Even the very boundary between conscious and unconscious—the so-called repression barrier—is understood, both developmentally and in the therapeutic situation, not as a fixed intrapsychic structure, but as a fluidly shifting property of an ongoing intersubjective system. (2000:149–150)

Stolorow's "contextual psychology" recognized the "constitutive role of relatedness in the making of all experience." Empowered with this insight, he argued that, "the Cartesian bifurcation is mended and inner and outer are seen to interweave seamlessly. We believe that this post-Cartesian vision will, in the new millennium, enable therapists and clients to continue exploring hitherto uncharted regions of intersubjective space" (2000:150). There remains at least one serious problem with charting the vast frontier of intersubjective space. From Freud's work we know that subjects may not know "the true" motivations behind their behavior; mysterious drives of the unconscious mind, illusions, fears, and self-deceptions make it impossible for the subject to understand and order his or her own psychology. As Stolorow himself pointed out (2000:150), "One is always organizing one's emotional and relational experiences so as to exclude whatever feels unacceptable, intolerable, or too dangerous in a particular context." If a client cannot access or observe their own subjectivity, if they simply exclude the emotions and experiences they cannot organize, what can a therapist learn about them by taking an intersubjective perspective?

From Looking-Glass to Mirror Neurons: Biological Foundations of Intersubjectivity

Lately, neuroscientists have discovered "the mirror neuron system," what some have claimed provides a biological substrate for intersubjective understanding or mind reading. A team of Italian neuroscientists including Giacomo Rizzolatti and Vittorio Gallese (Gallese et al. 1996; Rizzolatti 1992 et al.) serendipitously discovered that certain "monkey-see-monkey-do" neurons in the frontal lobe of a laboratory monkey's brain responded both when the animal performed an action (picking up a peanut) and when the monkey observed another individual perform the same action. In brief, the mirror effect describes the coordinated nervous excitement that can be detected and topographically localized in the brain of both a performer and an observer. Subsequent research has discovered the same phenomenon in human brains, leading many different commentators to assert that the mirror neuron system "blurs the distinction between first and third person activity" (Schilhab 2007:741), enabling observers to mentally access the intentions and experiences of others by watching their actions (Gallese and Goldman 1998; Gallese et al. 2004; Iacoboni et al. 1999, 2005), hearing others act (Kohler, et al. 2002; Rizzolatti and Craighero 2004), and observing others use language (Aziz-Zadeh et al. 2006; Fogassi and Ferrari 2007). Mirror neurons appear to help subjects directly take part in the emotional states of others (Enticott et al. 2008; Schulte-Rüther et al. 2007), join in their erotic excitement (Mouras et al. 2008), feel when they are touched (Keysers, et al. 2004), share in their sense of disgust (Wicker et al. 2003), and even commune with them spiritually (Hogue 2006).

Intentionality is commonly described as one of the defining attributes of subjectivity or consciousness (Luijpen 1960:92–93; Searle 1993: 9–10). Thus, the proposition that mirror neurons in an observer's brain are able to immediately inform themselves of an actor's intentions is nothing short of shocking. Iacoboni and his team focused on the role mirror neurons play in assessing the intentions of actors:

> Mary is grasping an apple. Why is she grasping it? Does she want to eat it, or give it to her brother, or maybe throw it away? The aim of the present study is to investigate the neural basis of *intention* understanding in this sense and, more specifically, the role played by the human mirror neuron system in this type of *intention understanding*. The term "*intention*" will be always used in this specific sense, to indicate the "why" of an action. (2005: 78–79; emphasis added)

"Mirror neurons," Rissolatti and others asserted, "provide a direct internal experience, and therefore understanding, of another person's act, *intention,* or emotion" (2006:56; emphasis added). They explained:

> Given that humans and monkeys are social species, it is not difficult to see the potential survival advantage of a mechanism, based on mirror neurons, that locks basic motor acts onto a larger motor semantic network, permitting the *direct and immediate comprehension* of others' behavior without complex cognitive machinery.

> In social life, however, understanding others' emotions is equally important. Indeed, emotion is often a key contextual element that signals the *intent* of an action. That is why we and other research groups have also been exploring whether the mirror system allows us to understand what others feel in addition to what they do. (2006:59–60; emphasis added)

The intention-reading power of mirror neurons perhaps has been most strongly asserted by Vittorio Gallese and Alvin Goldman, in an article "Mirror Neurons and the Simulation Theory of Mind-Reading." The two researchers claimed that mind-reading mirror neurons enable an observer to detect "another agent's goals and/or inner states" and help "him anticipate the agent's future actions" (1998:495–496). The "action-goal understanding" that mirror neurons support, as Gallese and Goldman concluded (1998:500), "constitutes a necessary phylogenetical stage within the evolutionary path leading to the fully developed mind-reading abilities of human beings."

The psychologist V.S. Ramachandran speculates that the discovery of mirror neurons is the "single most important 'unreported' (or at least, unpublicized) story of the decade. I predict that mirror neurons will do for psychology what DNA did for biology: they will provide a unifying framework and help explain a host of mental abilities that have hitherto remained mysterious and inaccessible to experiments" (2000). Ramachandran maintains that "mind-reading neurons" afford "a neural substrate for figuring out another person's intentions," and that they "may have played a crucial role in many quintessentially human abilities such as empathy, learning through *imitation* (rather than trial and error), and the rapid transmission of what we call 'culture'" (2007; italics original).

The mirror neuron system, as described by many of the neuroscientists cited here, seems to contribute to social order by enabling one mind to immediately grasp the intentions of another. Thanks to this newly discovered biological resource, it appears that an observer of Mary "grasping an apple" will know what she intends to do with it as quickly and clearly as Mary herself knows. As Iacoboni and his collaborators suggested (2005), we do not have to guess if she will eat it or throw it away, our neurons will know that she intends to give it to her brother. Sociologists have long appreciated the value of observing actors, but they have had trouble feeling perfectly secure about attributing meaning and intentions to the behavior of those they observe. If mirror neurons give observers the capacity to know the intentions of others just by watching them, will the interpretive methods of neuroscience soon replace those of ethnography? The central question becomes: is observing the same as understanding?

When it comes to research on mirror neurons, there is a common feature in the design of most experiments that is of crucial importance. Monkey or human, the subjects attached to brain-monitoring electrodes or being observed with functional Magnetic Resonance Imaging devices are expected to *perceive* another's action. They see an actor grasp or smell an object—and then the mirror neurons fire. They may watch a film clip of actors engaging in a practice, hear a sound that can easily be linked to a recognizable action, or they may hear someone speak—and then the mirror neurons fire. If mirror neurons can read the minds of others or permit access to another's intentions, why do they not fire independently of perceptions? If we can already read her mind, why do we need to wait until we see Mary grasp the apple before we know that she will do so and for what intention?

ASSESSING PERCEPTIONS AND MANAGING UNDERSTANDING

Whether they are neuroscientists or social scientists, it is worth noting that many of those who use the concept of intersubjectivity to suggest that participants in society have access to some form of common mind also point to the role communication plays in the genesis and development of both subjectivity and the negotiation of understanding between people (Aziz-Zadeh et al. 2006; Bradfield and Knight 2008). Participation in communication implies that one engages in conspicuous behavior that is as perceivable as, for instance, Mary grasping an apple. According to Berger and Luckmann,

> I speak as I think; so does my partner in the conversation. Both of us hear what each says at virtually the same instant, which makes possible a continuous, synchronized, reciprocal access to our two subjectivities, an intersubjective closeness in the face-to-face situation that no other sign system can duplicate. What is more, I hear *myself* as I speak; my own subjective meanings are made objectively and continuously available to me and *ipso facto* become "more real" to me. (1966:37–38)

The activity of saying and hearing has the potential to make subjectivity "more real" because communicative behavior and one's participation in it can be perceived. In order to think, it appears that one must "hear" oneself talking (Mead 1934:47). To gain insight into another's thoughts, one must hear the other speak. To understand the members of a group, ethnographers must observe collective behavior. For mirror neurons to access another's subjectivity, one must first perceive the other's actions. Whenever it comes to understanding the intentions of another, the *perception of behavior* seems to be the critical factor. This does not mean that subjectivity itself can be perceived; only that perception is a necessary, but not sufficient, condition for interpreting another subject's intentions.

The value of perception for gaining information about one's world and the intentions of other people cannot be underestimated. However, such self-constructed information must not be confused with that which is perceived. Heinz von Foerster mentions the "principle of undifferentiated coding" to make this point clear. The body's nerves may relay a greater or lesser intensity of localized excitement, but it is up to consciousness to code perceptions with differentiated meaning (von Foerster 1985:58). If one sees Mary grasp an apple, one has already privately coded the excitement of optical nerves—cells that are limited to registering energy in the form of light (Heider 1959). Perceiving only light, one may still expect to identify self-constructed unities such as Mary, apples, tables, and many other things. If one can see that Mary is grasping an apple to eat it and not to give it away, the difference must be visible in the symbolically coded form of her grasp. In other words, Mary and her observer must both have been culturally prepared to anticipate a variety of performance styles for grasping apples. For observers of meaningful grasps, the meaning of any particular type of grasp lies in its difference from other types of grasping.

Mary's thoughts, her behavior, and indeed every aspect of her world remain outside of every other observer's consciousness. Nonetheless, Mary may selectively perform behavior with her audience of observers in mind. If she believes her observers have been culturally prepared to make sense of different patterns of behavior, she may deliberately play to their expectations and perform her grasping in a way that strategically leads her audience to a particular assessment of her intentions. As ethologists Donald Owings and Eugene Morton described "the dynamic interplay" between participants in communication, Mary may be seen as an active "manager" of observers that she recognizes as active "assessors" of her behavior (1998:29). To manage others in this manner, Mary must be conscious that others perceive her and that they inform themselves with what they see. Niklas Luhmann described the way in which the "perception of perception" enables subjects to strategically manage the information processing of others:

> Perception is primarily a psychic acquisition of information, but it becomes a social phenomenon, that is, an articulation of double contingency, when one can perceive that one is perceived. In social situations ego can perceive that alter sees, and can perhaps also see what alter sees. Explicit communication can link onto this reflexive perceiving, thereby supplementing, clarifying, and delimiting, and it builds itself into this reflexive perceptual nexus because of course it depends on perception and on the perception of perception. (1995a:412)

Behavior that is perceived has no meaning until it is interpreted by an observer. However, both the performer and the observer may interpret what they perceive with the same cultured frames of symbolic reference, and both may know this of each other.

Despite their operational closure, psychic systems may be structurally coupled in the course of participating in communication. The structural and semantic resources of a social system (language, codes, symbolic generalizations, themes, programs) have the potential to inform autonomously thinking subjects because the resources may be perceived by subjects as anticipated, *socially organized forms of noise* and because the perception of such symbolically charged noise may itself be perceived (von Foerster 2003). Domenico Tosini described the subject's capacity to associate meaning with perceptions as follows:

> A property of the processes of consciousness is *symbolic perceiving*, which can be understood as the capacity to associate symbolic contents with operations located in the nervous system. Consequently one can conceptualize psychic systems (consciousnesses) as being systems whose operative closure consists of the production of symbolic perceptions referring to, and generated on the basis of, other symbolic perceptions. (2006:546)

For anything that is perceived to be coded with meaning, the subject must self-referentially imagine a symbolic connection that is not immediately available in the world of material things, bodies, brain cells, and energy waves. Mutual participation in communication provides subjects with the ability to expect one another to imagine approximately the same—*functionally equivalent*—symbolic connections between perceptions and meaning. As Luhmann maintained, "expectations acquire social relevance and thus suitability only if, on their part, they can be anticipated…Ego must be able to anticipate what alter anticipates of him to make his own anticipations and behavior agree with alter's anticipation" (1995a:303). There is no direct connection between alter and ego, no intersubjectivity or sharing of minds. Instead, operationally closed subjects mutually allow society to condition their own meaning-making activity. Aware that society presents a shared reference point, subjects turn to communication when they desire to assess and manage one another.

This sociological conception of how individuals consolidate their own private versions of reality appears as a basic proposition in the much-appreciated client-centered therapy developed by psychologist Carl Rogers. The "private world of the individual," Rogers wrote, "can only be known, in any genuine or complete sense, to the individual himself." "For social purposes," Rogers reasoned, "reality consists of those perceptions which have a high degree of commonality among various individuals. Thus this desk is 'real' because most people in our culture would have a perception of it which is very similar to my own" (1951:485). As Rogers talked and listened with a client in order to understand the latter's subjectivity, he presumed that "communication is at all times faulty and imperfect." Yet, despite its "cloudy fashion," "increasing communication gradually brings more of experience into the realm of awareness, and thus a more accurate and total picture" of the client's private world (1951:495–496).

SELF-REFERENCE AND INTEROBJECTIVITY

Constructivist Ernst von Glasersfeld emphasized the "essential and inescapable subjectivity of linguistic meaning" and the importance of trial and error in "the process of accommodating and tuning the meaning of words and linguistic expressions." Von Glasersfeld asserted,

> The basic elements from which an individual's conceptual structures are composed and the relations by means of which they are held together cannot be transferred from one language user to another, let alone from a proficient speaker to an infant. These building blocks must be abstracted from individual experience. And their interpersonal fit, which makes possible that which we call "communication," can arise only in the course of protracted interaction with others, through mutual orientation and adaptation...
>
> Understanding is a matter of fit rather than match. Put in the simplest way, to understand what someone has said or written means no less but also no more than to have built up a conceptual structure that, in the given context, appears to be compatible with the structure the speaker had in mind. And this compatibility, as a rule, manifests itself in no other way than that the receiver says and does nothing that contravenes the speaker's expectations. (2008:62)

Neither watching an actor perform a behavior, nor listening to the actor describe his or her intentions can achieve intersubjectivity between the actor and the observer. Observers must interpret the meaning of what they perceive with reference to their own cultured expectations. Yet, as they participate in society, subjects learn to expect that other subjects will expect them to attach certain "socially constructed" meanings to what they perceive. They learn that they cannot impose their private self-reference on socially significant gestures because the social world of meaning is, as Mead put it, "objective" and independent of their consciousness. No matter what participants intend, communication produces and constrains meaning according to its own self-reference. In Mead's words,

> Awareness or consciousness is not necessary for meaning in the process of social experience...The social process relates the responses of one individual to the gestures of another, as the meanings of the latter, and is thus responsible for the rise and existence of new objects in the social situation, objects dependent upon or constituted by these meanings. Meaning is thus not to be conceived, fundamentally, as a state of consciousness or as a set of organized relations existing or subsisting mentally outside the field of experience into which they enter; on the contrary, it should be conceived objectively, as having its existence entirely within this field itself. (1934:77)

The self-reference of communication conditions the way participating subjects connect what they can see, hear, and sense with meaning. Although no subject can directly perceive meaning, subjects may learn to anticipate how others will refer to "objective" social rules and conventions in order to associate what they perceive with meaning.

Alfred Schutz explained that participants in face-to-face situations understand the meaning of one another's behavior by interpreting what is immediately perceived in terms of a "We-relationship" that frames any given encounter. In his own formulation: "I can live in your subjective meaning-contexts only to the extent that I directly experience you within an actualized content-filled We-relationship" (1967:166). Like Mead, Schutz pointed to a socially prepared vocabulary of signs and behaviors that have "objective" meaning outside of any given interaction involving particular subjects. On the one hand, this objective frame of reference enables subjects to imagine the same symbolic connections between what they perceive actors do and the meaning of that behavior. "I get to your subjective meaning in the first place," Schutz wrote, "only by starting out with your spoken words as given and then by asking how you came to use those words" (1967:166). On the other hand, it is this same objective frame of reference that enables social scientists to interpret the meaning of the behavior they observe in the field. As Schutz argued: "*All social sciences are objective meaning-contexts of subjective meaning-contexts*... All scientific knowledge of the social world is indirect. It is knowledge of the world of contemporaries and the world of predecessors, never of the world of immediate social reality" (1967:241). Schutz concluded, therefore, that "it can hardly be the function [of the social sciences] to understand the subjective meaning of human action." Both ordinary interpreters of everyday behavior and social scientists grasp the meaning of what they see by "stipulating the typical and invariant subjective experiences" within an "objective meaning-configuration" (Schutz 1967:245). If we want to know Mary's intention when we see her grasp the apple, we must *not* explore her subjectivity. To the extent that understanding Mary is possible at all, her behavior must be interpreted within the objective context of the "We-relationship" that conditions her subjective experiences as well as our own. This *interobjective* meaning-configuration is the self-reference of society, the system of communication that presents itself on equal terms to all of its participating subjects.

SOCIETY, CONVENTIONS, AND THE ORGANIZATION OF UNDERSTANDING

The linguist Arie Verhagen, in his book, *Constructions of Intersubjectivity*, correctly asserted that "All language, when actively put to use, involves the coordination of one cognitive system with another" (2005:76). In Verhagen's appraisal of intersubjectivity, linguistic expressions are "primarily cues for making inferences" according to social conventions. He noted,

> Human languages—whether they have some hard-wired structural features (as generative linguists tend to believe) or not—are very basic systems of conventions: all words (links of sound and meaning) are conventional, and at least a substantial part of patterns of usage is conventional as well. Conventions are mutually shared rules—your behaving in a particular way (driving on a specific side of the road, making a specific sound in order to produce a specific effect) is conventional because its motivation only consists in your expectation that others will do the same in similar circumstances, as for exactly the same reason, namely that they expect others to do so. Therefore, knowledge of language is indeed mutually shared, and cannot be transmitted otherwise than by social learning. (2005:3)

The mirror neuron system discussed previously represents exactly the kind of "hard-wired structural feature" that some claim enables one subject to directly access another subject's intentions. Extending Verhagen's argument, neither mirror neurons nor any other biological mechanism can physically produce cognitive coordination between subjects, because such coordination results only when observers mutually (but autonomously!) interpret their own perceptions with reference to the same learned social conventions. Furthermore, neuroscientists must have learned the convention "grasping an apple" and also conventions about measuring neural activity before they can correlate the firing of neurons with perceiving someone perceiving another person grasping an apple.

Cognitive coordination is improbable because of the operative closure of psychic systems. However, because language appears in the world as a form of noise, it may be perceived by different subjects who also perceive that other subjects perceive it. In this sense, the nerve-stimulating power of coded noise (language) has the potential to structurally couple different psychic systems. The noise of communication captivates attention and invites subjects to recognize its organized variety, anticipate its redundancies, and make use of its conventions—whatever the subjects' own intentions may be. The search for intentions (those of others and one's own) can only begin when order, redundancy, and conventions provide frames for them.

Understanding another's subjective intentions or experiencing cognitive coordination depends, therefore, on a reciprocal awareness that one exists in the world with others. Though subjects cannot make an impression on one another as subjects, they can do so with their bodies. In his memorable discussion of how to interpret the intentions of a woodcutter, one who may be keeping warm, exercising, sublimating anger, or showing off his strength, Alfred Schutz observed: "we know nothing about our woodcutter except what we see before our eyes. By subjecting our own perceptions to interpretation, we know that we are in the presence of a fellow human being and that his bodily movements indicate he is engaged in an action which we recognize as that of cutting wood" (1967:114). The bodies of our "fellow human beings" catch our attention by stimulating our body's senses. When one sees Mary reaching for an apple, it is possible that one's own mirror neuron system works along with her body. In the same way that our optical nerves register light without interpreting what they see, mirror neurons may get excited without knowing why. Jacob Pierre (2008) convincingly argued that mirror neurons do not demonstrate mind-reading abilities. Instead, he suggested that mirror neurons may inversely help one predict another individual's future movements *under the condition that one has previously attributed intentions* to the actor. If perceptions of Mary lead one to imagine that she intends to eat an apple, for instance, one's mirror neurons might present a sympathetic representation of her anticipated motor commands, "an internal model of action," required to grasp and bite the fruit.

The embodiment of subjectivity, of physically being in a world that is shared by others whose eyes we perceive to be looking back at us, led Martin Heidegger to consider intersubjectivity (1996). For Heidegger, intersubjectivity appeared to suggest an impossible, hopelessly idealized subject-to-subject relationship in which neither subject views the other as an object or thing. We may choose to care for the other, imagining that the other faces the same existential struggles we face, but the other remains an external object. This condition of being-in-the-world-with-others is precisely what Schutz described as the "face-to-face situation" or the "We-relationship," in which subjects "share a community of space and a community of time" (1967:163–167).

Sharing space and time enables participants in society to potentially reach one another's senses with objective force. Communication leads to the possibility of cognitive coordination (or cognitive disturbance, as they case may be) between people because its resources are physically disturbing. On the one hand, as Verhagen emphasized, language works because its operating conventions are external to its users. To make sense together, participants in communication must learn to conform and adapt to the same rules of expression. On the other hand, communication can structurally couple different subjects because utterances take form as externalized energy that has a physical impact on bodies. Luhmann explained that structural couplings are effective because they have a basis in reality, in "a continuum of material or energy" that humans can and must depend on as they participate in communicative practices (1997a:102–103). Cognitive coordination through communication is possible because subjects are objects for one another. Each is embodied and, for this reason, open and vulnerable to the same physical irritations, whether or not their perceptions are socially organized as meaning. Peter Fuchs stressed that irritation is the "detector of structure," and writes that "structure is irritability" (2004:77).

Communication and Understanding

This chapter began by identifying the operational closure of consciousness as a source of uncertainty that hinders mutual understanding between people. If each thinks alone, in secret, and according to his or her own self-constructed terms, how is "our mind" produced? If all subjects think for themselves, what can explain the organization or association of different subjects? Intersubjectivity, as proposed by a diverse group of voices, was evaluated as a possible source of mutually accessible information or consciousness. Hegel's world spirit, Marx's class consciousness,

Durkheim's collective representations, and Gidding's consciousness of kind represent early attempts to articulate a transcendent, mutually available form of subjectivity. Mead and Cooley suggested that human intuition or sympathetic introspection could explain understanding between individuals in society, but they stopped short of claiming that "social consciousness" was anything more than an imaginary accomplishment of each actor. Interpretive sociologists such as Weber, Schutz, Goffman, Berger, and Luckmann came very close to asserting the existence of intersubjectivity, but each found a theoretical technique (ideal types, objectivations, objective meaning-contexts, frames) to retreat from the idiosyncratic intentions and endless complexity of actual subjects. The strongest claims of intersubjectivity's ordering power came from psychologists and neuroscientists, particularly those reacting to the discovery of the mirror neuron system, by Rizzolatti and Gallese. The mirror neuron system appears to be stimulated by one's perception of the behavior of another person. Mirror neurons get excited when we perceive other actors, but they cannot impart an understanding of subjectivity or intentions. Concluding this chapter, we return our attention to communication's contribution to *interobjective* understanding.

Mirror neurons are part of the human biological system, and therefore cannot inform meaning as it is self-referentially constructed by psychic and social systems. Making this point clear, when John Searle claimed that consciousness is a biological phenomenon, he carefully differentiated "lower level biological processes in the brain" from "higher level features of consciousness." Consciousness depends on brain cells, but it is not informed by biological mechanisms. Searle wrote:

> All the stimuli we receive from the external world are converted by the nervous system into one medium, namely, variable rates of neuron firings at synapses. And equally remarkably, these variable rates of neuron firings cause all of the color and variety of our conscious life. The smell of the flower, the sound of the symphony, the thoughts of theorems in Euclidian geometry—all are caused by lower level biological processes in the brain; and as far as we know, the crucial functional elements are neurons and synapses. (1993:6)

Neurons and synapses may indeed be the crucial functional elements of consciousness, but they cannot provide *meaning* to perceptions. One of the most enduring contributions of symbolic interactionism is the idea that the self arises only in the course of negotiating the meaning of symbols with others. Participation in communication is what enables a subject to inform itself with the meaning of what it perceives in the world. This social interaction takes place without violating the operational closure of biological, psychic, and social systems. The body's nervous system processes impulses, the psychic system processes thoughts, and the social system processes communication—all three operating with their own exclusive medium. If perceptions have meaning for an observer, it is because consciousness has attached symbolic meaning to them. This symbolic meaning may be accessible to other subjects only indirectly, interobjectively, as the self-reference of communication. Giddings, in his *Principles of Sociology,* already recognized that "by means of language the individual shares in the general fund (the social mind), which thus becomes for him an impersonal objective influence" (1896:132).

Adopting the words used by Alfred Schutz, understanding between subjects is possible because different psychic systems learn to condition their own "subjective meaning-contexts" with the "objective meaning-contexts" made available in the "We-relationship." With its abundant forms of objective meaning, the We-relationship confronts psychic systems as the self-reference of society. Luhmann asserted that as elements of communication reproduce themselves, they create "an emergent reality sui generis." The reality of communication, he pointed out, is society (1995a:408). Excluding the subjectivity of people from society, Luhmann argued that only communication communicates:

> In the same fashion as communication systems, systems of consciousness (and on their other side, brains and cells, etc…) are also operationally closed systems which cannot entertain contact with one another. There is no non-socially mediated communication from consciousness to consciousness, and there is no communication between the individual and society. Every

> adequately precise conception of communication excludes such possibilities. (1997a:105)

The emergent, objective reality of society is as external to individual subjects as the "real world" sources of their perceptions. Every experience of reality is fundamentally a subjective experience, and therefore every subjective experience is meaningless for society. Each subject senses the physical world with their own nerves, and each also makes sense of the social world on their own. Nonetheless, when subjects mutually inform and condition themselves with the same social conventions, cognitive coordination becomes as likely as two people shivering together on a cold day. Two different people can suffer from the same icy wind although they do not share the same nervous system. Along the same lines, "our minds" may attach the same meaning to Mary's grasping of an apple, without anyone sharing consciousness.

It is hard to imagine anyone being overly concerned with Mary's intentions regarding her apple, but it is easy to imagine how she could inform an audience of her intentions if she wished to do so. Her grasping the apple to throw it in anger, as a conspicuous performance, might be recognized as one selection among many other socially established types of grasping. Grinding her teeth and aiming narrowed eyes at her brother, Mary grasps the apple and raises it behind her head. Though her brother has no awareness of his trembling mirror neurons, he is already ducking behind his hands. The probability of reaching such an interobjective understanding depends on the social resources for representing meaning within the We-relationship. More than anything else, if Mary hopes her grasp will express a certain intention and not another, she must be able to expect the cultured expectations of her audience. Socially established expectations, regardless of their depth or complexity, have a cultural genealogy that can be identified and traced within the history of society. In his psychohistorical analysis of "our modern mind," Peter Gay traced social resources for representing and associating meaning to a particularly innovative era. Participants in modern society, he suggested, "are children of that quarter century, 1890 to 1914, crowded as it was with men of genius, and with radical innovations in literary and artistic techniques" (1973:10). The subjectivity of a genius, in the same manner as that of any other thinker, reveals its intentions only vicariously: through the interobjective self-reference of society. Language, spoken and written, is the indispensible resource with which observers of society may actively coordinate their understanding.

Chapter 4

Language

A Medium of Communication

The present chapter focuses attention on the symbolic medium of language, the evolution of its structures, and the problem of how it overcomes (or hides) the improbability of communication. This discussion will deepen our understanding of the distinction between communication and consciousness. It will also make clear why people may participate in communication, but monkeys, dolphins, babies, and other social noise makers cannot. As we have already explained, readers should bear in mind that, from the perspective of social system theory, participants in communication are cultured observers of differences. Participants in society synthesize meaning by observing constantly shifting selections of information, utterances, and understandings.

For the most part, it is language that gives communication the appearance of being something that can be taken for granted as normal and easy to reproduce. The everyday use of language helps sustain the illusion that operatively isolated minds can, from time to time, communicate. In an incisive article entitled, "The Improbability of Communication," Niklas

Luhmann (1981) described three discouraging obstacles that stack the odds against the success of any communication attempt. One must keep in mind that, from the perspective of social system theory, communication is society. As described in chapter 2, communication is always conceived as a unity of information, utterance, and understanding. Society is conceived as the unity of all communication. According to Luhmann, the first obstacle to communication results from the fact that two psychic systems cannot share the same thought or consciousness. If meaning can only appear within a specific context, and if context for each individual is constructed from thoughts or elements of consciousness supplied by his own memory, how can meaning ever be shared between different people? What could lead two operationally closed thought processing systems to make an identical observation when both have so many other things they could observe?

The second obstacle standing in the way of every communication is the difficulty of attracting and keeping the attention of those we would like to reach with our selected utterances. The media of communication must be observed in time and space. It is highly unlikely that your partner will always be at the right place at the right time, ready to do nothing but chat as soon as you indicate interest. Surrounded by so many other sounds and sights and smells, why would anyone take notice of you?

The third obstacle is that even when communication is understood, it may not elicit an anticipated or desired reply. Information may be rejected or it may only be returned with silence. You may understand a beggar who stops you on the street with the question, "Can you spare some change for a man in need?" Your understanding of his question does not require you to make a donation. It also doesn't guarantee that you will agree that the man really is in need, that your need for money is any less than his, or that you will stop and ask him if he would like to follow you home for dinner. It is always unlikely that one element of communication, even if understood, will ever lead to another. As Little Richard sang it in 1957, "You keep on knocking, but you can't come in."

Luhmann emphasized that these three improbabilities are not just obstacles preventing communication from successfully culminating in understanding; they also "function as thresholds of discouragement and lead to abstention from communication if the prospects for it are thought to be inauspicious" (1981:124). As Luhmann argued, the overwhelming odds against successful communication must be defeated if any social system is to emerge:

> There will be a tendency to abstain from communication when the prospects of reaching people and of meeting with understanding and success seem to be poor... The improbabilities of the process of communication and the way in which they can be surmounted and changed into probabilities govern the formation of social systems. The process of sociocultural evolution can therefore be viewed as the transformation and expansion of the conditions for effective communication on which society constructs its social systems; this is clearly not just a process of growth but one of selection and of determining what kinds of social system are feasible and what kinds have to be rejected as too improbable. (1981:124–125)

Explaining how society is an emergent order that "transforms the impossible into the possible and the improbable into the probable," Luhmann focused on the structural evolution of communication media (language, dissemination, and success media) and corresponding cultural innovations that serve to increase the organized complexity of social memory. "All in all," he wrote, "a pronounced trend toward greater differentiation and specialization is discernable and hence also a need to institutionalize the arbitrary to an ever-increasing extent... At each new and higher level of improbably probable communication achieved through improved technology, balance must be restored through new institutional expedients" (1981:127).

Communication Media

Social system theory asserts that the improbability of communication is effectively reduced by evolutionary achievements called media. With the concept of communication media, we have much more in mind than "mass media." For the development of his theory of media, Niklas Luhmann is indebted to the psychologist Fritz Heider. In his pioneering research on how humans make sense of what they perceive, Heider (1926) distinguished between "medium" and "thing," a pairing that Luhmann modifies as "medium" and "form." In Heider's work, media are "loosely coupled" elements that appear as things when taking form as "tight couplings" that may be perceived. The tight or strict couplings that give form (bestowing the status of a thing) to a medium are temporary, leave no permanent traces, and allow the medium to make itself available for future forms. Luhmann easily incorporated these insights into social system theory. All of the recognizable sounds of a language, for instance, are potentially available for a speaker to use. To form a meaningful word, the speaker must combine selective consonants and vowels, uttering a tight coupling that takes form within the acoustic medium of language. Letters from the alphabet, for instance, can be selectively combined to form words. In the economy, money is selectively connected with property in a way that sellers and buyers can both anticipate (in the form of a price). In education, grades can be selectively associated with students in a manner that makes sense to both teachers and students. This does not imply that a customer will purchase a car at a suggested price or that a student will be happy with a grade received. It also does not imply that a speaker or a writer will convince a listener or a reader to accept a particular piece of information. The point is that a tight coupling represents a contingency, a form within a medium that could have been different. Observers make meaning out of forms because they recognize the selectivity of a tight coupling and know that it could have been different. The twin constructs of medium and form are extremely important components of social system theory, and further attention will be devoted to them. At this point, we want to underscore that all communication media (language, disseminating, and success media) require cultured observers to perceive the occurrence of a form, to differentiate medium and form as an observation or operation (Luhmann 1997a:197, 1990a:166).

For each of the three mentioned improbabilities of communication—understanding, reaching, and success—we can find corresponding types of media. First, *the medium of language* makes it probable that understanding occurs, in spite of the fact that communication involves the participation of isolated psychic systems. These mental systems self-condition themselves with society to independently use neuro-physiological processes to discern coded information embedded in what their senses perceive. Second, *the media of dissemination* make it probable that communication reaches an absent audience of address-ees. Television, radio, newspaper, and the Internet are familiar media of dissemination. This category of media also includes writing and script in their most elementary forms, as well as books, musical scores, bank ledgers, diplomas, email, and memoirs.

As disseminating media evolve, they may strengthen the potential of communication to reach and involve more and more participants, even participants who are not co-present. When communication is limited to those who share time and space, it is typically bounded by narrow social controls, expectations, and cultural pressures. Advances in dissemination media invite communication to escape these bounds. However, as the possible range of communication increases, the chances that specific contributions will be accepted decreases. Correspondingly, the chances for misunderstanding, rejection, and conflict increase. For example, within a century after the Bible was first mass printed on presses, violent conflicts arose about the "right" interpretation of Holy Scripture.[1]

Success media, the third type of communication media, evolved to facilitate the acceptance or "success" of increasingly differentiated communication. Success media, such as power, money, love, truth, and art, take on very different structural and semantic characteristics. However, all forms of success media are meant to enhance the probability that communication attempts will be accepted and make a difference in society. Armed with power, money, love, truth, law, or some other medium of success, one may dare to launch social missiles that would otherwise have no chance of becoming information for their targeted observers. We will explore success media in detail in the next chapter.

Scholars, including sociologists, traditionally describe language as a system of signs (Robinson 2003). In addition, these signs are often described as a resource that can be "used" by members of a group once they learn how to handle the "rules" of the "system." Following this familiar line of reasoning, one may observe "simple languages" used by some kinds of animals, such as monkeys and dolphins. In a socio-biological or anthropological perspective, sign-use is thought to enhance cooperation within a species and, therefore, it addresses the needs of its members. According to such descriptions, language appears as a valuable tool in the natural struggle for survival.

Instead of stressing the importance of signs, rules, and tools for collective survival, social system theory describes the primary function of language as the selective specification of noise. What many researchers consider language-use among animals is indeed a highly sophisticated and sometimes surprisingly complex specification of noise. Language is built upon elements of significant and signifying sounds, but it adds something to these elements that separates the medium of language from other signs, sets of signs, rules of using signs, and so on. To clarify this certain added value, we will investigate the aforementioned complex forms of noise-specification that may be observed in the behavior of animals and, we believe, in the very first forms of human behavior. Understanding just how much animals are able to communicate will eventually help us highlight the specific difference that is provided and reproduced by the medium of language.

NOISE-SPECIFICATION IN ANIMAL AND HUMAN BEHAVIOR

We have no interest in debating whether or not animals can communicate. For us, there is no point in pursuing this question because we have already stipulated that only communication can communicate. Animals cannot communicate, humans cannot communicate, and computers cannot communicate. In the following pages we concentrate on describing some of the characteristics of animal vocalization, along with the mechanisms by which they are related to and integrated with behavior. Taking time to observe animal vocalization makes sense, as we agree with the ethologists Robert Seyfarth and Dorothy Cheney, that there is ample evidence that "non-human primate vocal development is broadly similar to human language development" (1997:269).

As we related in chapter 2, evolutionary biologists Donald Owings and Eugene Morton recently studied the use of vocal signals in connection with goal-oriented behavior and "regulatory problems:"

> Natural selection should favor mechanisms that are effective in minimizing the difference between preferred and actual states of the other's behavior. An animal can regulate the behavior of another animal either through direct physical manipulation or by communicating. Communicating with vocalizations is just one way of regulating social circumstances. Concurrent changes in posture, locomotion, and body orientation might, in a male California ground squirrel, for example, function together to keep another male away from an estrous female. (1998:10–11)

If an animal can discourage his competitor by producing a noise, he can avoid the risks associated with using physical force. Owings and Morton reported that "selection favors those able to use signals to manage the behavior of others in their own interests" (1998:11). Managing another's behavior, in terms of social system theory, begins with observing how the other self-referentially observes his own world; that is, how he makes his own meaningful indications. In other words, one's viability often entails observing another observer.

Let us consider some of the earliest interaction between a human mother and her child. In the beginning, there is a total dependence of the baby on the mother. Due to its relatively long postpartum development and "clean slate mentality," a human baby would certainly die if left alone (Crelin 1987:243). While mothers could choose to destroy or abandon their offspring, what facts or conditions seem to create the greater probability that a mother and child will form a lasting social bond, a bond that may serve as a basis upon which to build other associations?

At the moment of birth, a healthy baby emerges from the womb, crying and rooting for something to suck. It is hard to imagine that the mother, her breasts already producing milk, would resist allowing the baby to nurse. Babies are known to "worm" their own way up from their mother's belly to her nipples. With or without any instinctive drive to nurture, the exhausted mother is likely to appreciate the cessation of crying as the child latches on. As the baby simultaneously nurses and breathes air, the uterus begins to contract, milk production is encouraged, "mothering hormones" are released, and the mother's body begins its postpartum recuperation.

Owings and Morton (1998:62) noted that in many nonhuman primate species, infants produce a call, similar to a human baby's crying, that stimulates negative emotions in adult listeners and elicits maternal responses. They asserted that the crying of human babies seems to manage adult behavior by simulating infant respiratory distress, increasing the probability that adults will take an interest in alleviating the source of the baby's discomfort (1998:62). In the first act of breastfeeding, the mother and child physically comfort each other. The German word *stillen*, which means to breastfeed, literally means "to make quiet." Crying manages mothers and breastfeeding manages crying babies.

As she matures, the baby may begin to appreciate, from the experience of trial and error, that nursing alleviates hunger. Consequently, she is likely to develop a recurrent interest in her mother's whereabouts and in ways to attract her attention. When she feels hunger, she wails away like a siren, making solid use of her most basic mechanism of attracting attention. And what happens next? Anyone but her mother might run away to escape the noise. Her mother, however, feels the "let down" reflex in her breasts as the crying triggers an involuntary release of milk. Even if she ignores the leakage and runs in the opposite direction, how long will she bear the increasing pressure in her glands—how long before a mammary infection sets in? In short, both mother and child have compelling reasons to develop a lasting association.

Most every parent would probably agree that healthy human babies arrive with an incredible power to produce noise. How does this innate capacity to make such unappealing sound eventually turn into communication? For many months after birth, an infant's vocal tract physically resembles that of an adult chimpanzee more closely than it does that of the infant's mother (Crelin 1987:118). Thus, in a structural sense, the baby will have to wait two years before she can produce the vowels and consonants used in human language.

As sociologists, we want to take on the task of describing the conditions under which communication can increase in structural and semantic complexity. According to the biologist-turned-sociologist Helmut Plessner, a baby laughs and cries by nature, but "unnatural" social processes are needed to develop a child's language skills beyond the production of noise and the simple imitation of sounds (Plessner 1970). As a healthy child matures, crying is replaced by other communicative mechanisms for managing the behavior of parents: smiling, laughing, cooing, and other senseless signs of supplication. Seeking to manage their babies, mothers speaking many different languages develop a surprisingly similar "language" known as "motherese."

Schematic spectrograms documenting cross-cultural acoustic regularities of motherese suggest that infants may be naturally susceptible to management by certain patterns of vocal sound, long before they are able to understand any meaning conveyed by words (Fernald 1992). In a pragmatic sense, if motherese were not an inherently probable or successful tool for managing babies and mothers, its acoustic forms would surely not demonstrate such regularity. As mother and child learn to use different sounds and movements to manage each other, the child slowly learns how to sever utterances from noise. This improbable ability will support the child's struggle to observe and participate in all subsequent communication with companions. Muriel Saville-Troike described young children as "essentially participant-observers of communication, like small ethnographers, learning and inductively developing the rules of their speech community through processes of observation and interaction" (1989:221). This is precisely the kind of observation of communication that we described in chapter two. Mother and child pursue their own goals (viability) by observing how each observes communication.

William Graham Sumner depicted the emergence of folkways as a "natural force unconsciously set in operation" (1906:3–4). In the same manner, "without rational reflection or purpose," humans produce noise as they experiment with techniques for surviving and attaining their goals. To attract attention, noise does not require an arbitrarily or culturally defined referent. Noise is just what it is—something that speaks for itself, irritating and fascinating the ears. If a man screams in any country, people who hear will turn and look at him in wonder. Particularly for primates, communicative gestures, sounds, and facial expressions are tools for survival, effectively used against improbable odds to attract the valuable attention of others—inviting, arousing, warning, placating, or threatening, nuanced only by the context of the immediate situation.

Simply by making noise, humans can solve one major improbability of communication—the initial capture of another human's awareness. By repeatedly making the same cry, hoot, yelp, grunt, or other vocalization whenever a certain situation arises, older members of a social group teach younger members which sounds may be expected and in which situations. Thus, each species of primates has its own set of unique vocal calls and each produces specific sounds for different situations. For example, vervets, a common cat-sized species of monkey native to Africa, demonstrate the most sophisticated voice-based sign repertoire observed among animals (Diamond 1992:143; see also Seyfarth and Cheney 1997). Vervets evidently use a vocabulary of at least ten different significant sounds; allowing individuals to call the attention of other vervets to the presence of leopards, eagles, snakes, baboons, unfamiliar humans, and other significant referents. If a vervet cries out the sound for snake, for instance, others will immediately search the ground. If the call indicates an eagle, listeners will promptly search the sky. Making a connection to the discussion of observation in the first chapter, vervets learn that communication draws a distinction between calls and noise. Noise occurs on both sides of the distinction, but the society of vervets codes noise as differentiated calls that can *selectively* indicate a variety of specific threats. Similarly, a mother hears her child make noise before and after it learns to observe and use the distinction between language and noise.

Bottlenose dolphins offer another intriguing example of the systematic use of animal vocalization (Tyack and Sayigh 1997). Within a few days of birth, both wild and captive bottlenose dolphins begin to vocalize, producing whistles, chirps, and clicks. The sounds of infant dolphins are erratic and irregular until they begin to develop a consistent whistle during their first year. This "stereotyped" whistle, called a signature, is unique to the particular individual and remains stable for decades, even as individuals learn to make an increasing number of variable calls. Why would dolphins develop signature whistles? Apparently, they have a need to indicate the difference between individuals. Dolphins are social animals, engaging in a variety of short- and long-term relationships with different individuals over their lifetime. A mother, for example, typically spends several years with her calf—not just any calf. Adult males form coalitions and may be observed together 90% of the time for many years. Females also form enduring teams, often helping to protect one another's calves.

It seems that from a dolphin's perspective, one dolphin is not just as good as the next: identity (and difference!) matters. Signature whistles identify individual dolphins in a manner conducive to the marine environment. Down in the deep, where the nose cannot orient and visibility is limited to a few meters, dolphins use their signature whistles to indicate their identity and location over great distances. According to Peter Tyack and Laela Sayigh, bottlenose dolphins learn to "imitate the whistles of animals with which they share strong bonds, and the diversity of whistles in a dolphin's repertoire appears to be related to the size of its social network" (1997:230). Dolphins tend to match the acoustic character of their variant whistles to those made by their current companions, even while preserving their own invariant signature (Tyack and Sayigh 1997:225). In other words, a group of dolphins may develop similar whistles, despite the fact that each member of the group retains his own personal signature. This remarkable communicative behavior, known as vocal accommodation, may also be observed among other species, such as birds (Thorpe and North 1966) and humans. John J. Locke and Catherine Snow defined vocal accommodation as an evidently universal "tendency to adapt aspects of one's vocal production (including aspects as varied as pitch of voice, vowel production, intonation, accent, lexical choice, and grammatical features) to match or approach those of the conversational partner" (1997:283). Vocal accommodation has vital communicative importance because it is an improbable phenomenon: two randomly selected speakers cannot be expected to use their voice boxes in the same characteristic manner, even if they do speak the same language. Thus, vocal accommodation itself may communicate a speaker's social identity (see also Saville-Troike 1989:14, 200–204). According to Locke and Snow,

> Vocal accommodation is a useful social process precisely because so many aspects of vocal productions serve as social markers... Even without conveying any content through speech we convey considerable content through ways of speaking. Speech has been identified as socially marked for the personal features gender, age, geographic origin, social status, and sexual preference, as well as for situational features such as status of interlocutor and formality of setting. All these features which constitute markers of identity thus become as well targets for accommodation from interlocutors. (1997:283)

Vocalization and vocal accommodation helps vervets, dolphins, humans, and other animals attract attention, coordinate behavior, protect their group against attack, and maintain social boundaries, thereby increasing the chances for both individual and species survival. Producing differences within the media of sound and light, individuals capture each other's attention with noise, gesture, and movement. Individuality and belonging are represented by distinctive signatures and vocal accommodation.

But does the use of vocalization and vocal accommodation by members of a species imply the capacity for using language? It is tempting to cover all of our examples with the metaphor of "language," and to speak of "animal languages" (Diamond 1992:143). With some measured appreciation of the value and complexity of the calls of dolphins, monkeys, infants, and other animals, we can now clarify the higher qualities of language and its unique capacity to specify the social noises made by humans, as well as describe how language contributes to that recursive process we identify as communication.

The Voice Box: More Noise and More Meaning

Compared with other animals, primates, according to the classical anthropologist Kingsley Davis, possess the most elaborate mechanisms for articulating sound (1949:39). Davis noted that the "meanings" of the "anticipatory and expressive symbols" used by primates are not determined by instincts, but conditioned by experience and must be learned within a social setting. It seems plausible that our human ancestors approached one another in the same manner that contemporary primates do—displaying overt body postures, gestures, snorts, and calls. After all, politicians appealing to voters still use such primitive forms of drawing attention to themselves. Walking erect, with their noses far from the ground and their faces exposed to one another's eyes, early humans would have easily observed differences in anticipatory and expressive symbols. Davis noted, however, that the communicative mechanisms of primates do not demonstrate the uniquely human capacity for what he calls, "arbitrary symbolic communication:"

> Anticipatory and expressive symbols, by virtue of their intrinsic connections with their meanings, are not pure symbols. They cannot be used, therefore, to depict situations that do not exist as a part of the current situation. An animal may simulate anger in order to bluff his companion, but he cannot describe to his companion another situation in which bluffing occurred. He may by his manifestation of fear indicate to his fellows that an enemy is approaching, but when no enemy is near he cannot describe what an enemy is like and teach his fellows what the appropriate behavior would be. (1949:42)

According to Davis, only a set of arbitrary symbols can handle absent or hypothetical situations. The handling of absence or hypothetical presence is an absolutely essential prerequisite for the emergence of what we call "reference," and—building upon reference—meaning and culture in general. Without transcending what is immediate and actual for all participants, culture and transferable knowledge would not be possible.

"It is this system (of arbitrary symbols)," Davis maintained, that "enables man to transmit attitudes and knowledge from one generation to the next, that makes possible the elaborate conditioning of the anthropoid bio-social traits" (1949:44). As we explained in chapter 2, culture transmits the imaginary code that selectively and arbitrarily associates sounds with things and events in the world.

The decisive difference between simple vocalization or vocal accommodation and the medium of language lies in the absence of a distinct referent, in the lack of a concrete object and a concrete attitude towards a concrete object. Only because of this absence can we distinguish the "saying" from the "meaning" of what is said. Vocalization demonstrates the tight coupling of sound and reaction—the immediate and unhampered connection of sound with an attitude and its corresponding behavior. There is literally no time for doubt or deliberation, and therefore no time for recursion. For the members of those species that accommodate their behavior with the help of vocal sound, the sound and the information are the same. In the loose couplings available in language, however, objects and attitudes may be selectively related or positioned with reference to other things. The relations are not inherent in the sounds. Therefore, the saying and its intention, the sound and its meaning, the utterance and the information must be distinguished—not only analytically, but by every participant in verbal communication.

If we understand the being of the world as what is perceivable within the world, then in communication, language transcends the world of being. It opens up the realm of "the possible:" what could have been possible, what is possible elsewhere, and what is possible in the future. Kingsley Davis clearly described the valuable capacity of "arbitrary symbolic communication" to separate itself from the world at hand, referencing absent objects and situations in the past, present, and future. Language indeed has this power, but in focusing on single words and their culturally defined meanings, Davis did not adequately explain how language works. He stopped short of appreciating a considerable point that resolves the mystery of language as a medium that specifies noise. Davis' analysis focused on significant single words, but in so doing he missed the actual elements of language. The form of language consists not of single sounds or single words, but of sentences: groupings of significant sounds. Language does indeed reference absent entities—be they absent objects or absent attitudes towards objects. But even the simplest sentence must do more than reference something.

It is very important to note that a meaningful sentence always references and associates at least two different elements of meaning, thereby building self-reference. Thus, each sentence used in language may be simplified according to the model: "A is B." For example, a sentence may reference and associate a man and his height: "Dave is 5 feet, 9 inches tall." It may reference a place and associate its amenities: "Ventura is a good place to surf." Adding complexity, a sentence may reference a person and associate her emotional ties to another: "She loves you, ya, ya, ya." Without delving into cumbersome linguistic models or matters of propositional logic, we can use the model sentence, "A is B," to help describe the process of specifying noise with language. In contrast to animal and early human vocalization and vocal accommodation, in which one sound or expressive behavior may represent an object or evoke a certain attitude towards a present and perceivable object, language involves at least two significant sounds that are related to each other. Furthermore, the task of relating sound A to sound B is completely self-referential: the constructed relationship has no equivalent, model, or clues in the perceivable environment. The connection between "A is B" must be actively produced internally, within the psychic systems of both (or all) participants in communication. Whenever people speak, the procedures of relating the uttered As and Bs are constitutively absent, if we define absence here as "not part of the perceivable environment." When communication successfully leads one to recognize a sentence as information, understanding the "A is B" relation happens in one's mind, not out in one's environment. This is also the case when both A and B are singularly perceivable, as in "This chair (A) is red (B)." In the moment of speaking and listening, the "being-red of that chair" has to be produced and reproduced in the minds of the speaker and listener.

We may now draw three central conclusions about the medium of language. First, language refers to sentences, not just highly sophisticated single sounds or words. Sentences are the forms of language, whereas words establish the medium of language (Luhmann 1997a:219–220). The forms and the medium of language make each other possible. Without words there can be no sentences, and words have no meaning outside of sentences. In contrast to other theories, we suggest reserving the term language for this specific relation between medium (words) and forms (sentences). The following two points, which will lead us back to communication and consciousness, refer exclusively to the sentence-type of noise-specification and are not valid for single word or sound vocalizations.

Secondly, language determines its own meaning (referents) with a precision like no other form of noise-specification. Language can do this because relations between sounds and their meaning are underdetermined. The meaning of the referent, A, can be suspended in time as long as the communication process is willing and able to explore all possible Bs that might be associated with it. Meaningful connections may be made between A and possible Bs independently of any "real-time" or "present" relationship. In contrast, whenever a vervet calls, "Here comes an eagle," the meaning of the call is strictly bound to an actual event and context—the immediate presence of an eagle is required. If the "eagle alarm" is sounded without the presence of its referent, the call will have only produced noise. Participating in communication, human conversants may talk about the "ice hockey goal of the year" again and again, despite the fact that the referenced event, the goal, lasted only a fraction of a second. Long after an event is over, it can continue to make a difference as a theme within communication. Similarly, communication may reference events expected to occur in the future. With great anticipation, hockey fans can discuss a future game, debating possible outcomes and contingencies to the smallest detail. People stake money on which team they believe will win a future game—a form of communication that requires an event to be temporally distant. Language transcends that which is perceivable at any given moment, and therefore opens up the possibility for communication to reference what used to be, what might be, and even what may never be present. Meaning can be elaborated independently from ongoing processes in the world, as long as the process of communication focuses on that meaning. Thus, the meaning of verbal sounds can reach a precision that cannot be found in any other medium.

Thirdly, language establishes the crucial possibility of negation. "The essential business of language," Bertrand Russell observed, "is to assert or deny facts" (1961:x). We would add that an assertion is meaningful precisely because it could have been a denial, and vice versa. With the vocalizations of animals, every expression of rejection is positively accomplished by a certain corresponding behavior. The vervet's call is exactly what it is: the utterance is the information. Only in language can we separate the two, accepting the utterance and negating the information. "The possibility of rejection," Luhmann suggested, "is necessarily built into the communication process:"

> Starting from this, we can define an elemental event of communication as the smallest unit that can be negated. This is meant not in the logical sense, but in the sense of communicative practice. Every proposition, every demand opens up many possibilities for negation: not this but that, not this way, not now, and so forth. These possibilities remain open as meaning references as long as ego has not reacted to them. The utterance itself only offers a selection. Reaction is what terminates communication, and only then can one tell what has emerged as a unit. This is why communication cannot be understood as action, especially if one asks about the ultimate, unanalyzable unit. (Luhmann 1995a:154)

Using language, we may negate something in order to establish something else in its place: "You don't accept my proposal? I am ready for a better suggestion!" It is verbal negation that allows communication (and all its participants) to build up and maintain distance from the "rest of the world," to de-couple its own processes from physical, biological, and psychological processes and events. It is verbal negation that enacts the recursivity of communication: we can take anything for granted as long as we can correct our assumptions at a later time, whenever necessary. Last but not least, it is the potential to say "no" that indemnifies the individual's risk of participating in communication, which would otherwise force inescapable bindings.

Precision, negation, and replacement work hand in hand and are mutually dependent on each other. None of these features can be found in sentence-less forms of noise-specification. What we describe as non-verbal communication (gestures, facial expressions, postures, and onomatopoeia of all forms) is not the foundation of communication. When it comes to the complexity of social situations, including the mutual perception of participants, language is the essential medium. Only language can establish the expectation of clarifying the situation and selecting between alternative courses of action. Non-verbal gestures can only "comment," punctuate, and sometimes counter-play on such expectations. Furthermore, only language renders the difference between information and utterance, as well as the possibility to clearly refer to this difference ("What do you mean by saying 'A is B'?"). Thus, we may conclude that only language enables communication. The three elements of communication—information, utterance, and understanding—rely on language in order to gain their identity and closure. Language alone provides the necessary distinctions required for the self-referential reproduction of communication.

Anything that can be negated, and treated as a negation within communication, may contribute to the autopoiesis of communication. We must emphasize that only statements can be negated, as when "A is B" is replaced with "A is non-B." Communication can dissolve itself into such units, depending on the precision of formulations and of negators. But communication cannot dissolve itself beyond the base level of statements. It cannot, for instance, negate the letter "c" in the word "communication," nor can it dissolve sound waves and the noise produced by voice boxes and registered by eardrums. Furthermore, communication cannot negate thoughts and imaginations—only the expression of such thoughts and imaginations. Communication establishes itself in the difference between what is stated and what is not stated (Luhmann and Fuchs 1989:7). Language works within the medium of words, producing meaning in the form of sentences, which may or may not be uttered and which may or may not be negated.

Meaning in Language and Consciousness

How does the form of noise-specification called "language" make a difference on consciousness and the mind? The mind is continually occupied with sensual perceptions delivered by the nervous system of the body (Luhmann 2000a:5–8). All the sounds offered by language remain in the realm of the perceivable. But, as we suggested above, the "A-to-B" relationship specified in every sentence has—as a relationship—no meaning that would or could be immediately apparent or immediately validated by perception. In other words, referencing the "A-to-B" form creates no recognizable redundancy in the rest of the world. Such redundancies must first be constructed, and, in the case of additional interest and further investigation, detected. Language, although perceivable by itself, loosens the relationship of the mind with its perceptions. Loosening does not mean that connections between mind and perception are suspended or abolished. Rather, language loosens by introducing minor or major degrees of freedom for the mind, permitting the mind to reach beyond immediate perception as it produces meaningful thoughts, fantasies, and imaginations.

Making sense of acute physical pain seems to provide a clear case of a tight relationship between the mind and body, when sensations are not only present but very difficult to deny. Nevertheless, even in this case, the linguistic indication, "I feel pain" (with A = "I" and B = "feel pain"), evokes the possibility of its negation, "I feel no pain." This loosening through language might lead to further mental associations, such as memories of more comfortable situations, hope for medical help, and even possible pleasure or future satisfaction through pain. The form of language urges and thereby enables the mind to modify what it perceives; to enrich its perceptions (including its fantasies) by mere modalities, by mere possibilities, by memories and expectations, and so on. In short, language brings the possibility of meaning into the world of immediate perceptions.

One might object: What about the crying and calling vervets? Did we not state that a specific call (A) denotes a specific piece of information, (B) "Attention, eagle arriving?" Of course, accustomed as we are to the medium and forms of language, we (not the vervets) can describe (for ourselves) the vervet's situation and the relationship carried by the call as a form of meaning. The vervets, however, can only handle the warning for caution as a unit, not as a difference. There is an immediate transformation of detecting noise into a state of caution. If animals demonstrate something akin to communication, it is the rudimentary and fragmentary flashing up and immediate collapse of meaning, with no lasting differences and without a sense for the experience of making meaning. Vervets do not have any calls that can describe the unity and difference of their calls.

Returning to our focus on communication and human consciousness, both of these systems operate permanently and exclusively within the form of meaning. The production of thoughts depends on suggestions of meaningful references of whatsoever might be perceivable or imaginable. The same holds for the production of (understandable) utterances. No thought and no utterance will ever transcend the form of meaning. "Thus meaning," Luhmann attested, "is an unnegatable category" (1995a:62). We might touch meaninglessness in a moment of unspeakable pain or black despair—but in the next moment we either attach indications and meaning to the experience, or the touch will have faded without leaving any difference in further operations of the mind and the social system.

We have described both communication and consciousness as systems that operate in the medium of meaning. We have also tied the emergent production of meaning to language. With regard to socio-cultural evolution, we may now conclude that language does not merely establish the difference between animal and human behavior in terms of relative intelligence. Within the world of sensual perceptions, language introduces meaning, communication, and consciousness at the same time. As an inseparable quartet, language, meaning, communication, and consciousness are recursively dependent on each other. Only as a whole can this quartet establish the difference of meaning within the world of mere perceptions, non-meaning, sheer indifference, or nature. Without language, there would be no meaning, no communication, and no consciousness. To make this point clear, Luhmann contrasted first- and second-order forms of communication:

> What language makes possible, namely, the communication of comprehensible sentences, are thus second-order forms—forms in the medium of what a consciousness is able to process in terms of perception. It is only at this level that a social system can differentiate itself on the sole basis of self-produced communication. Only on this level do we arrive at an autopoiesis of society which—under minimal restrictions of perceptibility and therefore with the participation of consciousness—organizes itself into what constitutes, for it, a medium, but what, for consciousness, is already form. Only at this level of the construction of order can forms be aptly observed with the help of the distinction between speaking and silence. Accordingly, communication is set up in such a way that it fascinates consciousness by the use of first-level forms and carries it away by the use of second-level forms. Especially significant communication is, in the first instance, formed in such a rhapsodic manner, using the acoustic medium, and only its secondary encoding in phonetic writing makes possible a certain distance. (1994:29)

The distinction between humans and animals is not complex enough to appreciate the transition in socio-cultural evolution that was brought about through the emergence of language. For this we would need a theory of the co-evolution of biological, psychic, and social systems. We do not yet dispose over such a theory: the relevant distinction between communication and consciousness must be more fully developed. Fortunately, archaeological physiology offers some provocative information about the evolution of the biological prerequisites of language: the evolution of the so-called "voice box." In his book, *The Third Chimpanzee*, the physiologist Jared Diamond expressed strong support for the theory that it was the perfection of the voice box, that laid "the anatomical basis of modern language:"

> The structure of the larynx, tongue, and associated muscles give us fine control over spoken sounds. Like a Swiss watch, all of whose many parts have to be well designed for the watch to keep time at all, our vocal tract depends on the precise functioning of many structures and muscles. Chimps are thought to be physically incapable of producing several of the commonest human vowels. (1992:55)

Crelin (1987:5) noted that the evolution of the voice box "occurred only during the last million years of a period of time consisting of many millions of years of human ancestral evolution."

According to Diamond (1999:38), the earliest Homo sapiens who were our ancestors differed from us in skeletal details, had significantly smaller brains, and were "grossly different" from us in their cultural and behavioral forms. A half a million years ago, while still confined to Africa and Eurasia, humans had developed very primitive stone tools and were able to use and control fire. The early Homo sapiens did not produce art, bone tools, wheels, textiles, or boats. Based on ample skeletal evidence, the human populations of Africa, Europe, and western Asia between 130,000 and 40,000 years ago proceeded to diverge from each other. In Europe and Asia, Neanderthals began to make small improvements to their tools, bury their dead, and care for their sick. Diamond (1999:39) asserted that the Homo neanderthalensis, like the earliest Homo sapiens in Africa, "rank as less than fully human." He summarized the cultural products of the earliest humans as follows:

> Although those Africans of 100,000 years ago had more modern skeletons than did their Neanderthal contemporaries, they made essentially the same crude tools… still lacking standardized shapes. They had no preserved art. To judge from the bone evidence of the animal species on which they preyed, their hunting skills were unimpressive and mainly directed at easy-to-kill, not-at-all dangerous animals. They were not yet in the business of slaughtering buffalo, pigs, and other dangerous prey. They

> couldn't even catch fish: their sites immediately on the seacoast lack fish bones and fishhooks. (Diamond 1999:39)

In Diamond's opinion, the voice box was the physiological innovation that made possible our "Great Leap Forward," when "human history at last took off around 50,000 years ago" (1999:39). Archeological evidence from sites in East Africa, the Near East, and Europe suggests that Cro-Magnon humans, as Diamond puts it, "were as modern in their minds as they were in their skeletons" (1999:40):

> Cro-Magnon garbage heaps yield not only stone tools but also tools of bone... Tools were produced in diverse and distinctive shapes so modern that their functions as needles, awls, engraving tools, and so on are obvious to us... Recognizable multipiece weapons at Cro-Magnon sites include harpoons, spear-throwers, and eventually bows and arrows... Those efficient means of killing at a distance permitted the hunting of such dangerous prey as rhinos and elephants, while the invention of rope for nets, lines, and snares allowed the addition of fish and birds to our diet. Remains of houses and sewn clothing testify to a greatly improved ability to survive in cold climates, and remains of jewelry and carefully buried skeletons indicate revolutionary aesthetic and spiritual developments. (Diamond 1999:39)

Armed with their voice boxes, language skills, and weapons, the Cro-Magnons took over Europe some 40,000 years ago, quickly annihilating the Neanderthals "who had been evolving as the sole occupants of Europe for hundreds of thousands of years" (Diamond 1999:40). Floating boats, hurling flint missiles, and wrapping up their mostly hairless bodies in textiles, the Cro-Magnons moved on to occupy Australia and New Guinea, Siberia, and the Americas—rounding out their presence on all five continents.

Why was the voice box such an asset to the Cro-Magnons? The voice box is a structural device that can be used to produce an amazing number of finely differentiated sounds. With practice, the possible units of sound can be repeatedly and constantly distinguished from each other. Working together with a complementary receptive device (the eardrum), the voice box provides the physiological precondition for the possibility of differentiating the "A/B" form of sentences, and thus opens up the medium of language. In turn, the actual realization of that possibility depends on the dynamic interplay of communication and consciousness, the only vehicle for attaching meaning and understanding to sound. Deeply appreciating its structural complexity, Crelin described the anatomy and physiology of the vocal tract and explained how it produces uniquely human sounds:

> The vocal tract part of the upper respiratory system consists of the oral cavity, or mouth; the oropharynx and laryngopharynx, or throat; and the larynx, or voice box... Phonation begins with the larynx, which produces the fundamental frequency of a sound. The larynx functions as a sphincter to regulate the air flow into and out of the trachea... The regulated opening of the larynx... is actually a slit-like gap in a dense sheet of fibroelastic tissue... the free edges of the fibroelastic membrane are known as vocal ligaments (vocal cords)... The vocal ligaments are attached anteriorly to the thyroid cartilage; posteriorly, each is attached to the process of an arytenoid cartilage... Thus, the numerous intrinsic muscles that act directly or indirectly upon the thyroid and arytenoid cartilages cause the vocal folds to be separated and loose or taut, or approximated and loose or taut, to variable degrees. As air rushes from the lungs and respiratory tubes to pass between the approximated vocal folds, it forces the folds to separate and then slam together at rapid intervals. This releases puffs, packets, or waves of compressed air that create sound waves. These, in turn, cause the eardrum of a person to vibrate. (1987:13; see also Pinker 1995:163–169)

As they are forced outward, the fundamental frequencies of sound waves produced in the voice box are "filtered" into a formant frequency by the supralaryngeal part of the vocal tract. Vowel sounds are created when the flow of air is unobstructed, while consonants are formed by the sudden release of obstructed air. The lips, teeth, and tongue may contribute final modifications of sound waves. Social system theory shares a similar profound interest in the evolution of the structures and technical processes involved in producing and controlling "the noise" of communication, from the voice box to the pen and computer keyboard.

"In an evolutionary context," Luhmann asserted, "language [*Sprache*] is an extremely improbable kind of noise that, because of its improbability, is highly able to attract attention and lends itself to highly complex possibilities of specification" (1997a:110). When specific forms of speech are differentiated from noise and given meaning in a social context, a language emerges that can, with luck, adequately bridge the gap between individuals who are trying to understand each other. According to Luhmann,

> Language possesses its very own form. As a form with two sides it emerges from the difference between sound [*Laut*] and meaning [*Sinn*]...Spoken communication is the processing of meaning within the medium of sound...In order to construct the difference between medium and form within language itself, the medial substrate of language, the difference between sound and meaning, must be underspecified. Without underspecification there would be nothing left to say because everything would already have been said. This problem is solved with the difference between words and sentences. Words are also constellations of sounds with meaning; but they do not determine the sentences that they can produce when combined...With the help of language, one may say something that has never been said before. (1997a:213–215)

Producing and specifying connections between vocal sounds and meanings, mutterers became utterers and yellers became tellers. Without a voice box, Neanderthals made a much more limited variety of sounds, sounds that were also difficult to combine in meaningful ways. Imagine having to construct a sentence from a yell, a hoot, a wind suck, and a buzz. Consider how much easier it would be to build a language using vowels and consonants, making use of changes in tone, pitch, and volume. As Diamond asserted:

> It's easy to appreciate how a tiny change in anatomy resulting in capacity for speech would produce a huge change in behavior.

> With language, it takes only a few seconds to communicate the message, "Turn sharp right at the fourth tree and drive the male antelope toward the reddish boulder, where I'll hide to spear it." Without language, that message could not be communicated at all… two protohumans could not brainstorm together about how to devise a better tool, or about what a cave painting might mean. Without language, even one protohuman would have had difficulty thinking out for himself or herself how to devise a better tool. (1992:55)

The important point is that Cro-Magnons had the structural means to produce very many different vocal noises. The first arbitrary semantic associations of vocal sounds with referents were most likely built on top of a much simpler learned "vocabulary" of calls, gestures, and postures (perhaps similar to the calls of vervets). Just as modern speakers use gestures and body language to punctuate their vocal sounds, Cro-Magnons probably punctuated their gestures with vocal noises. The voice box, because its possible emissions were both highly diverse and controllable, eventually became the signaling mechanism of choice. The structural development of the voice box enabled an increasingly complex specification of human noise that could produce the differentiated forms of language, the basic medium of communication. As John Dewey pointed out, it is participation in language that establishes one's social identity:

> To learn to be human is to develop through the give-and-take of communication an effective sense of being an individually distinctive member of a community; one who understands and appreciates its beliefs, desires and methods, and who contributes to a further conversion of organic powers into human resources and values. (1927:154)

Coded Noise: The Muse of Society

Practicing over generations, humans learned to control vocal sounds, reproducing them at will, and reproducing them within an increasingly complicated semantic context. It is easy to imagine an older child vocalizing to get his mother's attention, and then pointing to his or her own mouth. If inclined to help, the mother would wonder if the child wanted food, water, or sympathy for a toothache. In the interaction between mother and child, the most expedient solution to such a problem of ambiguity would be the systematic manipulation of vocal noise into coded signs of difference: Food or water? Nuts or berries? Almonds or pistachios?

The crying baby, mentioned above, demonstrates highly impressive noise making abilities, but lacks the capacity for the "A-is-B" specification of noise. Drawing a conclusion, we must now deny the baby the capacity for language. More critically, we would also have to deny its capacity for communication and consciousness! Interestingly, Crelin (1987:5) noted "the mature form of the human vocal tract occurs very late, actually long after birth." The insight that babies cannot communicate, should not surprise us any more: neither can mothers and fathers. Only communication communicates! But what about the baby's cry, is it not an obvious sign of personal need? Is it not self-evident that the baby is crying because it needs help? As a matter of fact, this is exactly the intended meaning parents tend to *attribute* to the cry! They must assume the presence of intentions and expressions, and act on their own assumptions. For the baby, crying is just crying and the world is just the way it is; in the beginning, there is no memory or registry of differences. By inferring intentions to the baby's cry and testing the validity of alternative inferences through corresponding behavior (holding, feeding, moving, introducing objects, etc.), parents offer regularities and redundancies in the perceivable world, providing the baby with the opportunity to differentiate its own noise in response. The baby learns that what is perceivable can be subjected to modification through callings. This learning juncture, often described in theories of human development, occurs when the quartet of language, meaning, communication, and consciousness emerges within the baby's natural (that is to say, meaningless) world. According to Peter Fuchs (2002a; 2002b), communication continues to flood the neural and nervous system of the baby until consciousness, which in the beginning is merely supposed, slowly but surely asserts itself and can announce in a manner parents can understand, "Here I am." Only from this point on can the quartet of language, meaning, consciousness, and communication play out its normal game, a game so quickly taken for granted that we tend to forget about the improbability of understanding.

From the moment of an individual's first thoughtful observation, language provides a crucial structural link between consciousness and communication. We mentioned above that language (and only language) generates and implements the possibility of a determined negation. Within language, the positive and the negative form of propositions and statements are easily exchanged. The crucial part of nearly every verbal interplay consists in adding or omitting a "no" here, an "I don't believe" there, or an "I don't want to" to some other phrase.

Even while fully participating in the flow of communication, consciousness must constantly manage the difference between positive and negative statements. Because language loosens the tight coupling of meaning and perception, the self-referential management of affirmation and negation is the only way to detect potential errors, mistakes, and deceptions (both intentional and unintentional). The "well-socialized" and "reasonable" mind is supposed to tell what is the case, and to tell what is not the case; to tell what has happened, and to tell what has not happened; to tell what ought to be achieved and what ought to be avoided, and so on. Furthermore, it is supposed to tell or at least to assess with high certainty whether the utterances of other people tell what is the case or if they belong to fantasy. One must determine whether one's partners in discourse are right and authoritative or if they are wrong and in need of correction. One must be able to evaluate whether they follow honest intentions or if they are to be opposed, ignored, repeated, or censored. The binary codification of language, the distinct affirmations and distinct negations, is "the muse of society" (Luhmann 1997a:225). To take up and keep a meaningful address—to be included in complex communication—one must accept the burden of drawing distinctions between affirmation and negation and understanding the results of such operations.

Language, we repeat, is nothing more than a particular form of noise. The forms of language are sentences and its medium is the words. Sentences establish the "A-to-B" relationship between specifically coded sounds. This is the starting point of meaning and of the detachment or liberation of meaning-based operating systems from mere nature—that is, from what is immediately apparent. Thus, the emergence of language requires and catalyzes the co-evolution of consciousness and communication as two distinct systems; both of which operate exclusively within the form of meaning. On the one hand, language, for cultured observers, makes understanding within the realm of meaning possible and probable. And it does this although the perceptions and the thoughts of all participants remain strictly isolated and separated! On the other hand, the apparent ease with which language overcomes the difficulties of reaching and understanding others makes it hard for us to re-discover and analyze the operative closure of consciousness and communication. Everything seems to be connected with everything—but it is only language that binds everything with everything, and thereby feeds the necessary illusion of direct connections and contacts.

This chapter has introduced the deep interest of social system theory in understanding the historically changing structures that produce the noise required to perceive and observe communication. As the structures (voice boxes, smoke signals, semaphores, word processors, etc.) of language evolve, the capacity to differentiate noise and selectively associate meaning increase. We have emphasized that both animals and humans use sound to manage participants in their social worlds, but language is an exclusively human form of noise produced only by society. In the next chapter, we extend our focus to other forms of communication and analyze how they encourage the success of society along very different channels.

ENDNOTE

1. The printing press was invented in 1455 and came into general use in the 16th century (compare Eisenstein 1979). In the beginning of the 17th century, Europe experienced an intense and long-lasting religious war in which the interpretation of scripture played a significant role.

Chapter 5

Symbolically Generalized Media

Encouraging the Success of Communication

As Alfred North Whitehead explained, the human mind functions symbolically when one component of experience is associated with another component (1959). Within the mind of an observer, a thought, a feeling, or a thing may be connected to a specific meaning with an imaginary bond. Ernst Cassirer reasoned that symbolic forms mediate between one's perceptions of the world and one's understanding of what reality means. If the world of possible references is imaginary and is mediated by symbols that exist only in the mind, how do individuals use communication to assess and manage one another? Answering this question requires that we augment a philosophical or psychological account of how individual psychic systems create symbols with a sociological understanding of how symbols are created by society and made available for collective use by participants. In this chapter, we will explain how communication gives meaning to actual social practices by culturing participants to expect each other to reference the same symbolic distinctions. These symbolic distinctions, which we will identify as communication media, become meaningful when observers observe each other using them selectively, choosing one possibility rather than another.

When we investigate the use of symbols by psychic systems, it becomes evident that different individuals seem to imagine an overwhelming number of the same symbols. This is not a surprise, of course, but it is a condition that needs to be explained. Sociologists traditionally refer to the socialization process to make clear how individuals learn to adopt and "internalize" the use of their community's symbols. According to George Herbert Mead, one's intelligence and self-awareness emerge in the course of participation in communication. Mead wrote, "Only in terms of gestures as significant symbols is the existence of mind or intelligence possible; for only in terms of gestures which are significant symbols can thinking—which is simply an internalized or implicit conversation of the individual with himself by means of such gestures—take place" (1934:47). From the perspective of social system theory, because psychic systems process all of their mental operations as closed systems, there can be no internalization of society. Psychic and social systems are autonomous and have their own boundaries, so we must distinguish the significant symbols used by each. In the course of socializing themselves, participants in society discipline and coordinate their private vocabulary of symbols with the symbols they learn to anticipate and associate with society. When they have properly acquired sufficient culture, individuals "understand" the same symbolic references: they use the same imaginary associations when speaking to themselves and to others *and they expect that others will do the same*. Using communication to manage others depends on our ability to expect the expectations of other observers.

In the previous chapter, we described how language symbolically references meaning by specifying noise. The capacity to perceive and strategically modify the same sources of environmental noise sets up the possibility that noise can be culturally organized into meaningful symbols. To grasp how participants in communication assess and manage each other by specifying noise, we need to differentiate several problems. First, there is the problem of raw perception. How can a noise maker expect that noises will be perceived by a listener? The second problem pertains to expecting particular transitions. How can a noise maker anticipate that a listener will make a transition between a certain noise and a certain symbolic meaning? Noise must be effectively coded so that participants in communication can expect one another to make the same associations. Let's assume that a father tells a child, "Look both ways before you cross the street." Parents know that speaking to a child does not necessarily indicate the presence of an active listener. The child might be deaf. The ability to hear the acoustic waves of a language does not necessarily lead to understanding. The child might not speak English. He or she might not be old enough or mentally prepared to hear advice or even notice the presence of another person. Under such conditions, the parent may not be able to reach the child with the intended meaning because the noise of language was either not perceived or the transition from noise to meaning was not made. The use of coded noise to assess and manage others also brings up a third problem, the issue of connecting a sequence of symbolic references. A listener may be successfully reached by noise and may correctly associate coded noise with symbolic meaning; but the listener may still fail to connect the association with any other meaning, let alone the one intended by the speaker. The child "understands" the advice and carefully look up and down ("both ways") before walking into traffic. This final problem, managing the way participants in communication reciprocally associate their own individual symbolic references, is perhaps the most fascinating issue for sociology. How does communication increase the likelihood that participants—each autonomously processing his or her own thoughts—will make the same connections between sequential observations of meaning? This issue of establishing connectivity gains enormous importance when participants attempt to negotiate the meaning of vitally important highly contested matters such as money, truth, power, and love. Establishing connectivity in the flow of communication, as we will see, depends on society's ability to condition and coordinate the expectations of participants.

SELECTIVITY AND THE PROBLEM OF DOUBLE CONTINGENCY

Everyone knows that money and love should never be mistaken! Money and power should also be clearly distinguished, just as much as power and truth or love and power. Nonetheless, basic familiarity with the dramas of real life, classic novels, contemporary pulp fiction, and Hollywood films leads one to appreciate just how often such phenomena are confused. There is the lady who pretends to love only to charm a wealthy husband; the man who pays a mistress for sex, and finds himself falling hopelessly in love; the politician who uses his power to make some secret extra cash; the husband who forces his devoted wife into humiliating submission; the government that strategically keeps scientific discoveries secret; the professor who cloaks her favorite ideology with academic claims of truth, and so on. Confusion between money, love, power, and truth provides substance for the best dramas and stories about life, at least for authors who want to easily attract an audience. And yet such phenomena must be clearly distinguished. If they were impossible to tell apart, or if there were no differences between them, we could not entertain ourselves with the countless possibilities of confusion. More than that, we could not anticipate how our experiences and actions might influence the experiences and behavior of those around us. In social system theory, money, love, and a number of other phenomena are brought together under a single sociological construct: that of symbolically generalized media of communication.

Symbolically generalized media help communication succeed by focusing selectivity and restricting meaning to the confines of a selected "central code," a specific, previously established, two-sided distinction (Luhmann 1997a:359). Media such as money, truth, power, and love have evolved to help process communication within distinct, functionally specific, simultaneously operating societal systems. This "division of labor in society" increases the expectation that a given understanding will be accepted and lead to further communication. For this reason, they are also called *success media*. Luhmann explained,

> Every communication *stakes a claim*. At the very least, it demands time and attention. In addition, every utterance, however circumspect, expresses the expectation that it will be accepted, and this expectation of success can be greatly reinforced, especially with the help of symbolically generalized media of communication... Anyone who initiates communication or expands the thematic repertoire of a system with new elements would do well to keep in mind communication's severe demands if he wants to make sure of the communication's chances: *he is extending the system's boundaries*. (1995a:195)

Many of communication's themes are redundant or familiar and their frequent reappearance can be anticipated by participants. Symbolically generalized media have emerged with the redundancy of such themes. New themes, as Luhmann suggested, challenge participants to spontaneously make sense of unfamiliar claims, references, and possible connections.

Talcott Parsons pioneered the concept of symbolically generalized media, which he viewed as part of the "shared symbolic system" that integrates modern society. Parsons conceived of a basic problem for which shared symbols appeared to be the solution: the problem of double contingency. How can any social order be established if two members of the smallest social group, alter and ego, cannot anticipate one another's actions? Parsons argued that double contingency is inherent in every interaction:

> On the one hand, ego's gratifications are contingent on his selection among available alternatives. But in turn, alter's reaction will be contingent on ego's selection and will result from a complementary selection on alter's part. Because of this double contingency, communication, which is the precondition of cultural patterns, could not exist without both generalization from the particularity of the specific situations (which are never identical for ego and alter) and *stability* of meaning which can only be assured by "conventions" observed by both parties. (1951:16)

At every moment, alter and ego can individually decide whether or not they will conform to or deviate from social conventions, but they assume risks when deviating from cultural "ways of orienting" communication. As long as they mutually guide their selections with the norms of a shared symbolic system, cultural patterns help alter and ego stabilize meaning and order their reciprocal expectations. Following Parsons, we agree that symbolically generalized media are cultural conventions that invest a small number of alternatives with meaning, while bracketing out all other possibilities. One may walk into a store and assume that items are available for sale. The owner of the store assumes that shoppers have money available. Both the shopper and the storeowner determine their own decisions, but each expects that the other will orient their selections with respect to price. If the shopper is reluctant, the seller may decide to lower the price. If the owner is reluctant, one may offer to pay more. There may be no final sale, but both participants in the interaction understand what it would have required to make the deal. In a generalized manner that escapes particular circumstances, prices symbolize economic value for a buyer, a seller, and others who may observe the interaction. By manipulating the amount of money specified by a price, a seller selectively uses a symbolic medium of communication to manage the decision of a buyer. Of course, the buyer may confess to not really having any money to spend. The seller may decide to give an item away for free. Each participant could decide to behave in countless ways that would not be meaningful within an economic interaction. When buyer and seller are serious about seeking a mutually gratifying transaction, however, observing the symbolic medium of money serves the private interests of both.

Contemporary social system theory adopts Parsons' symbolically generalized media, but the construct has been altered in two important ways. First, the focus is no longer on the problem of coordinating actions or in reciprocally "gratifying" the interests of actors. Instead, the concern is to understand how symbolically generalized media increase the probability that communication will establish its own organization and connectivity. We will return to this point. Second, Parsons was convinced that symbolically generalized media effectively solve the problem of double contingency because members of society share unifying values or normative consensus. Within its own sphere, morality may indeed condition the acceptance of communication with the force of universal dictates for "good behavior," but absolute notions of goodness cannot guarantee the success of communication in an increasingly differentiated modern society. Within relatively undifferentiated premodern societies, individual participants condition their own experience and behavior with a common construction of reality that provides general assumptions about nature and an ontological fundament for social order. Selecting good behavior is a matter of conforming to universally applicable, commonly acknowledged dictates. In the "folk society," everyone knows what everyone should do in every situation (Redfield 1947). In a functionally differentiated society, however, societal systems such as the economy, politics, or education process communicative selections without reference to an overarching moral distinction between good and bad. Demonstrating what evolutionary biologists describe as adaptive polymorphism, communication reproduces itself by simultaneously processing meaning with reference to many different functional distinctions, including morality. As Luhmann put it, symbolically generalized media are the "functional equivalents" of a general morality (1997a:317–318). What morality might accomplish for society, other distinctions can do with a greater probability of success, efficiency, and peace. That social order does not depend on normative consensus among actors is one of social system theory's most controversial and insightful assertions.

Societal systems such as the economy, politics, education, art, and the family operate as observers with reference to a selected problem of communication, and each one attempts to connect its own sequential operations with the help of a different symbolically generalized medium. As an observer of its own kind, a societal system draws a distinction that includes one side as its preferred or "positive" alternative, connectable to its past operations, and the other "negative" side as its reflexive value. The reflexive value stimulates questions about how an indicated "no connection" may still lead to a connection in a subsequent communicative operation. The structural distinctions drawn by societal systems are, as Luhmann asserted, "symbolic abbreviations representing highly complex expectational situations." As Luhmann explained:

> A structureless chaos would be absolutely insecure; only that would be secure. Basically, the concepts of security and insecurity have no meaning for such a state. Through the differentiation of expectational structures, this state is replaced by a combinatory interplay of relatively secure or insecure positive and negative expectations. (1995a:307)

The problem of double contingency, as Parsons correctly maintained, is that it creates a condition of uncertainty and "social disorder." If both make selections independently, partners in communication cannot predict how each will behave. Luhmann tied the experience of contingency, however, to the possibility of organizing behavioral expectations within a system. He wrote:

> What the experience of contingency achieves is the constitution and opening up of chance for conditioning functions within the system, thus, the transformation of chance into structural probabilities... Wherever one encounters another under the condition of reciprocally experienced double contingency, a continuation of contact can be achieved only by agreement between selective behavioral determinations, and this is achieved only by forming systems... The connection between double contingency and system formation... enables the evolution of specifically social orders—so that evolution means only the construction and destruction of structured orderings on an emergent level of reality. (1995a:120–121)

Ego determines which side of a system's structured distinction should be indicated by conditioning his selection with alter's choice. "She did this instead of that, so I will do that instead of this." Participants in society are observers observing observers. Ego and alter are not only observing each other, they are observing each other observing a social system that reduces complexity by setting up specific dilemmas. Uncertainty always exists because our own choices depend on the choices that will be made by others. Harrison White and his colleagues pointed out that there is no need to limit the appreciation of contingency to dyadic processes involving alter and ego. Instead, they suggested analyzing conditions of "multiple contingency" within the "mesh of interrelations" characterizing networks (White et al. 2007:546). Nonetheless, the uncertainty resulting from double and even multiple contingency may be reduced when observers mutually observe a societal system that establishes highly constrained symbolic parameters in which choices are invested with meaning (for an empirical account, see Lee 2009). The principle of limiting possibilities with selective constraints was first suggested by the anthropologist Alexander A. Goldenweiser, who used it to explain the development of culture (1913). Ego and alter confront each other as decision makers who make choices from a socially constructed set of possibilities. They both recognize the alternatives made available by a certain symbolic distinction, and this sets up the chance that they will seek to anticipate and manage one another's choices.

Success Media and Forms of Connectivity

The various types of communication media are devoted to overcoming the probability that communication will fail. *Language* overcomes the improbability that speakers and listeners (or writers and readers) will achieve social understanding despite the operational closure of each individual's consciousness. From the earliest technologies for writing, to telegraphs, telephones, cassette tapes, and the latest electronic media, *dissemination media* overcome the improbability that information will reach an audience. Dissemination media decentralize control over communication and free participants in society from the need to be physically close and synchronized in time. For their part, *success media* increase the chances that an understanding will result in the acceptance of a claim, an offered form of meaning. Success is achieved when the recipient of an utterance takes the selective content of the communication, the information, as the premise for his own behavior, connecting the selection to other selections and thereby increasing their selectivity

(Luhmann 1981). Accepting a selection as the premise for one's own behavior can lead to a number of different kinds of connections: following a command, experiencing a work of art, reflecting on a text, or forming an opinion about the meaning of a gesture. When it would make more sense to expect rejection, symbolically generalized media make this kind of acceptance probable by reducing complexity, excluding all but two meaningful possibilities. Success media work, when they work, by presenting an unresolved dilemma and inviting observing systems to internally inform themselves about determining a meaningful choice.

It makes sense to draw explicit attention to the three conceptual aspects of symbolically generalized media. Success media are *symbolic* in that they employ the self-reference of society to connect a signifier with a meaning that must be imagined by a network of cultured (socialized) observers. They are *generalized* in the sense that their meaning appears and reappears as a familiar redundancy in different operations. A success medium is generalized because it appears as a rule, as Ludwig Wittgenstein might have put it, that observers can apply in more than just one situation. Along with other communication media, language and dissemination media, success media are described as *media* for an important reason. It is not because they are symbols that mediate, in Cassirer's sense, between perception and reality. It is also not because they mediate meaning between participants in communication, as Parsons appears to have meant. Niklas Luhmann used the word media with reference to a theory of form and thing developed by the psychologist of perception, Fritz Heider (1926), whose work was introduced in chapter 4. The concept of a medium, contrasted to a thing, is grounded in the distinction between loosely and strictly coupled elements. Adapting Heider's notion, Luhmann maintained (1995a:197) that the loose, pliable, formable structure of a medium enables temporary forms (things) to appear for observers. As an illustration, Heider mentioned the impressions made by one's feet when walking on sand at the beach. The sand is the (loosely coupled) medium in which footsteps take temporary (tightly coupled) form. When a speaker uses language, a listener may construct words as things within the perceivable acoustic medium of sound. When a writer composes a poem on paper, verses take form by becoming things within the optical medium of light. When it comes to a symbolically generalized medium, scientists may observe forms of truth (astronomy, mathematics, physics, sociology) within the medium of knowledge, which includes everything that might be studied (astrology, numerology, metaphysics, humanism). Individuals may observe different forms of love for their parents, children, romantic partners, and neighbors.

Cultured observers are aware that the same medium may give rise to different forms; any actualized tight coupling is a contingency that is neither necessary nor impossible. Thus, when used in a particular social operation, the meaning of a success medium may be described as the difference between an actualized form and its medium. For Luhmann, "meaning can be meaning only as the difference between what is actual at any moment and a horizon of possibilities, every actualization always also leads to a virtualization of the potentialities that could be connected up with it" (1995a:65). Alter and ego know that any given strict coupling could have been different. The price of a new pair of shoes could be higher or lower than what the price tag declares. The awareness that every form is a contingency is why alter and ego can both understand that it is meaningful when one form of a medium appears rather than another. Forms are selective arrangements of elements, tight couplings, that refer observers to other observers who make conditioned decisions about actualizing a medium's possibilities.

To understand the logic of success media, it is central that the notion of "success" not be attached to the factual consequences of communication—that is, to people successfully attaining their goals. Communication is exclusively interested in its own success, in its own continuation. Much more importantly, media situate success as a prospect and in doing so motivate attempts to solicit the acceptance of meaningful proposals. When it comes to an individual's own learning and socialization, it is certainly true that the anticipation of success is dependent on the past achievement of results. One who has never experienced successful communication will not believe in the prospect of future success. Nonetheless, depending on the medium, the connection between practical results and motivations may be quite flexible. Consider, for instance, that the prospect of receiving love is rarely given up—even in the face of repeated, miserable disappointments—and it may continue to motivate new proposals over a lifetime. Although it often promises very limited chances of success, communication in society can still maintain the high motivation of many contenders. Lottery and roulette players provide a simple example of this resistance to frustration. In politics we may think of the promise of government by the people for the people, while in reality only very few positions are charged with political power. Thus, with regard to communication and "success," one should primarily think about motivation and only secondarily about participants getting what they desire. Concerning the factual consequences of communication, one must remember that success always brings along its second side as a shadow: success and failure belong together. Failure returns awareness back to the improbability of communication; enriched, however, from a certain experience. The experience of failure may become the subject of reflection and facilitate learning, conditioning expectations and motivating a fresh attempt to make a connection.

Causality and Motivation

How do success media bring about the prospect of acceptance when one ought to expect rejection? The underlying mechanism depends on applying the concept of *causality*, connecting a cause with an effect, to relations between people. The meaning of causality is normally learned at a young age, as described most notably by Jean Piaget in his studies of early childhood development (1960:128–231). We can abbreviate the general form of causality in simple terms as: A causes B. From practical experience with his immediate surroundings, a child learns two meaningful principles. First, the child learns that its own behavior can make changes in the environment. He throws a ball and a window shatters. In this case, throwing the ball (A) caused the breaking of the window (B). Second, the child learns that changes in his environment can make changes in himself, modifying his own state. A dog barks and the child experiences fear. In this case, the dog's bark (A) lies in the environment and causes the internal effect of fear (B).

Causality is a familiar concept in traditional sociology, traceable to the discipline's Enlightenment-era origins in scientific positivism. Positivist researchers strictly apply the scientific method to measure associations between variables, stressing the value of collecting empirical data that demonstrate links between causes and effects. Social system theory values the concept of causality, but "radically refuses positivism" (Luhmann 2005:163). Developing what Luhmann described as a "hermeneutic" or interpretive approach to causality, social system theory applies the concept to explain how social practices or processes use binary structures to reduce complexity. If the social practices that actually take place are interpreted as causes and effects, then that which occurs empirically may be viewed as a selective actualization—a decision that reduces complexity. Participants in communication see not only what is actualized practice by practice, but they also consider the horizon of possibilities from which one practice was realized. If we construct what happens as an effect, we can search for possible causes. If what happens is identified as a cause, we may look for possible effects. Either way, causes and effects are interpreted as selective actualizations that entail meaning by making reference to a symbolic distinction. The causality of social practices is a consequence of the informative power of symbolic distinctions, not of scientific laws or natural forces. Participants in society actualize their own choices—this option and not that one—because they have learned to recognize the choices of others as contingent selections.

Causality is an extremely important scheme for ordering meaning, especially with reference to things in the world. Without making use of causality as an ordering schema, we could not cook, ride a bike, ring a doorbell, or play with a ball at recess. With reference to the world of things—what we can do with them and what they can do to us—causality appears to be a matter of technique. Adding a bit more thyme to the sauce will make it taste delicious. A few drops of oil will make the squeaky door hinge quiet. With regard to communication, success media are available frames or patterns for organizing relations between causes and effects within society. For many centuries, a technique for communication was taught in schools to the upper classes: the art of rhetoric. Rhetoricians advised that if one merely spoke with the proper words (cause A) and presented the appropriate illustrations, one would achieve the desired effect (B) on one's audience. Words were assumed to possess the power to directly cause desired effects, a belief that has been more or less abandoned with modernity. Of course, though it may seem ridiculous, some people still attempt to manipulate others with magic words, as this believer in "love spells" testified on her Web page:

> Kyra is the newest spell caster I have tried. I only ordered one *Love Spell* from her, but let's just say, the results were amazing! I actually don't want to put too many details here, in case my new FIANCÉ finds this site. Yes, I said fiancé! :) We are getting married next spring! I cannot believe that I have found a love so true and so amazing and so perfect! This was by far the best $100 I have ever spent!!! Much better than the free stuff I was trying on my own! (Shadowcoven 2006)

Casting love spells is an old fashioned technique for harnessing the power of words to cause effects, but modern lovers still use language to manage their partners. The great difference, perhaps, is that those who cast love spells do not appreciate their targeted partner's ability to refuse the proposal. Success media only function for observers who acknowledge that their partners in communication autonomously determine their own selections.

The success media of communication take causality for granted and are founded on the meaning scheme that selectively links causes with their effects and effects with their causes. Participants in communication would learn to ignore each other if they could not take for granted that their own selections establish conditioning premises for the selections of others. If I assume my choice will make a difference on my companion, then I can meaningfully evaluate what my choice should be in order to cause the desired effect on my companion. Symbolically generalized media use the principle of causality with a specific limitation: causality is restricted to the experience and behavior of persons. This primary reference to people, observers of distinctions, differentiates success media from technique and causal relations between things.

All success media are constructed in such a way that they use meaning to causally connect the experience or behavior of one person (ego) with the experience or behavior of another person (alter): an experience causes behavior, behavior causes an experience, an experience causes an experience, or a behavior causes a behavior. This matrix of possible causal relationships observed by two people, Alter and Ego, along with their respective success medium, is depicted in figure 5.1.

FIGURE 5.1. Matrix of causal relationships between experience and behavior with success medium (translated and modified from Luhmann 1997a:336).

Ego / Alter	Experience	Behavior
Experience	Alter's experience motivates Ego's experience: *Truth*	Alter's experience motivates Ego's behavior: *Love*
Behavior	Alter's behavior motivates Ego's experience: *Money*	Alter's behavior motivates Ego's behavior: *Power*

As a reminder, social system theory does not presuppose the existence of people. Alter and ego are cultured observers of a communication process that may or may not, according to its own self-reference, bestow "personhood" on its participants. In other words, the theory emphasizes that being human does not make one a person, it does not guarantee an address from which to participate in society. As Luhmann wrote:

> A person is constituted for the sake of ordering behavioral expectations that can be fulfilled by her and her alone. One can be a person for oneself and for others. Being a person requires that one draws and binds expectations to oneself with the help of one's psychic system and body, including expectations about oneself with regard to others. The more expectations and the more different types of them that are individualized in this way, the more complex the person. (1995a:315)

Persons, in Luhmann's view, are "collages of expectations, functioning as points of reference for further selections within the system" (1995a:127). Thus, when we speak of ego and alter using communication to manage and assess one another, we presuppose that communication has already raised reciprocal expectations of personhood. One expectation, for instance, is that when ego and alter participate in communication, each one will be observable as a combination of both ego and alter (Luhmann 1997a:333).

In this special form, in this restriction to the experience and behavior of persons, causality takes on the name of *motivation*. Money motivates the giving away of valuable goods. Power motivates the accomplishment of annoying and difficult tasks. Love motivates the tolerance of extremely unusual behavior. Truth motivates the acceptance of preposterous explanations. All symbolically generalized media motivate, establishing the wonderful source of the fascinating confusions that we mentioned at the beginning of this chapter. Success media differentiate themselves, however, in how they connect the experience and behavior of ego and alter. Observing how motivation is cultured by society so that it can meaningfully connect alter and ego is a fundamental goal in sociological research. When ego and alter understand each other's motivations, they do so by observing with the help of society's distinctions.

Money, Truth, Power, and Love

Before we examine other general features and functions of success media, we will describe the basic structure of motivation with regard to the four examples depicted in figure 5.1—the cases of money, truth, power, and love. Niklas Luhmann's own matrix included the additional media of property, law, art, and values. In the interest of providing an accessible introduction to the theory of symbolically generalized media, we will limit our analysis to only four (for additional study, see Luhmann 1979, 1989, 1990b, 1995c, 2000c, 2000d).

Traditional theories of *money* tend to mention three functions: exchange, calculation, and the conservation of value. This point of view is not false and is not at all disputed by the theory of success media. Nonetheless, in order to serve any of these functions, money must first function as a medium of communication. In other words, if money is not recognized and accepted as a medium of communication, it cannot serve any other functions; it could not be exchanged, counted, or saved. As a success medium, money must connect the experience and behavior of different people in a very specific manner. Beginning with the metaphor of exchange, it would appear as if money links a series of distinct actions. One person takes, another gives, and each of these actions is doubled. One accepts the goods while the other accepts the money; one gives the money while the other gives the goods. This constellation of four actions raises the question of how to attribute causality, how to assign causality in a manner that could be recognized as the same to all participants. To pursue this matter, we must broaden our perspective beyond a buyer and a seller and include additional participants in society, all those who simply observe the exchange from a distance and who do not discern their own advantages. Performing no active role in the exchange, the "virtual" participants enable or allow the taking of something valuable. Each "third participant" experiences how another person gains an advantage, without gaining any part of it themselves. On the level of motivation, money connects this passive observation with property-accumulating behavior. Thus, the behavior of alter motivates all observing egos to restrict themselves to mere experience, preventing active intervention. This is the form of causality attributed to money in figure 5.1. As money is paid, the one who takes it provides the chance that in the next moment a different person can seek the same advantage. When a bicycle is listed for auction on eBay, for example, we can imagine that the seller certainly hopes that a buyer will be motivated to purchase the item. Beyond that hope, however, the seller also anticipates that other people will eventually be motivated to accept the cash he receives from the person who bought the bike. Money builds peace between the individual participants in economic communication, discouraging hostilities that would destroy the possibility of future competition and trade. Drug dealers, gun smugglers, and pimps appreciate a peaceful business climate as much as any respected member of the chamber of commerce. For all participants in the economy, the value of money is tied to its continued circulation as an observable success medium. Taking a sociological perspective, we may observe a buyer and seller observing the economy as a social system that meaningfully connects their current exchange with economic operations made in the past and operations that might be made in the future.

Operation by operation, money connects motivations and makes improbable acceptance probable. What exactly is improbable when it comes to economic communication? Ego cannot expect alter to simply surrender a valuable good, service, or quantity of money. Property is scarce and many people would like to possess it. Money serves as a medium for observing the difference between scarcity and abundance. Economic operations involve payments of money, exchanges of one scarce item for another. Using the symbolically generalized medium of money, every kind of property or service can be assessed and expressed with a number. It is improbable that alter will part with a treasured item, but that probability increases with the number of dollars offered by ego. As ego and alter negotiate their transaction using the same medium of money, social system theory observes the improbable continuation of economic communication.

Whenever money and property change hands, a tight coupling of motivations (a specific causal relationship) dissolves and the medium of money is ready for recirculation in future economic operations.

The success medium of *truth* does not connect behavior and experience, but rather experience with experience. Science's distinction between true and false plays a role only with regard to a claim of new knowledge. It is concerned with surprises, with gaining information one did not expect or already know. Those who receive new information are, of course, cast in the role of being uninformed. It seems unlikely that anyone would gladly accept the label of being unapprised. Thus, at the very least, communication must invent some reformulation that makes it possible to differentiate the state of "not knowing" from the state of being uninformed. This invention is called truth. Truth symbolizes the pure conveyance of one person's experience to the experience of a second person. There can be no claims of superior intelligence for the one who informs, because personal traits, motives, or interests do not play a role in establishing the meaning of truth. For this reason one says that truths are "discovered," and not that they are created or produced. Discovering truth certainly involves action. One must equip oneself with the proper tools, travel far into the bush, and search hard for some new fossil or unknown mineral. The same hard work is required before and after every experiment conducted in a laboratory. The researcher must write down and publish a description of exactly what was done and precisely what was experienced—requiring still more active exertion. Nonetheless, in the scientific texts that are written and read, the symbol of truth always signals that anyone who follows the same procedures would experience the same results. The scientist realizes that his discovery is something exceptional (otherwise it would not be new knowledge), but that it is not exceptional as a personal experience. It is not true because one wants it to be so or because it reflects on oneself in a special way; it is true because anyone who puts himself in the same situation of discovery would have the same experience.

Following the lead of Max Weber, sociologists tend to agree with philosophers of science that scientific knowledge cannot determine ultimate truth (2004). Research on the professional behavior of scientists provides ample evidence that they are seldom dedicated to the pursuit of knowledge and the uncompromised experience of discovery (Latour and Woolgar 1979). On the contrary, the dominating interests appear to be for career advancement and establishing a personal reputation. This awareness, however, does not surprise us since we know that the central problem is to make improbable acceptance probable. The observations of philosophers and sociologists of knowledge merely bring back into view the symptoms of improbability that are hidden by the institutionalization of communication referring to truth (academic societies, universities, publishers). What is decisive for us is not whether or not truth exists, or whether scientists are mainly interested in uncovering truth or advancing their careers, but what encourages people to make or accept unusual assertions about knowledge. Visual observation seems to guarantee that the earth is flat, so why would anyone risk making the claim that it is round? Sociologists of knowledge must also ask themselves if their assertion, that scientists are more concerned about themselves than truth, is itself true or merely communication that they themselves find interesting. Why would anyone accept the cynical claim that scientists are not really concerned with truth? Only truth itself, making a difference as a success medium, can encourage such an acceptance.

Each success medium is internally complex and deserves its own focused analysis. However, one important point applies to all symbolically generalized media: each success medium must differentiate itself from all of the others. We do not accept a claim of knowledge as truth because someone received money for his research, or because a few students fell in love with the professor who made the claim, or because the scientist is protected by aggressive corporate lawyers. We accept a claim of truth if it is described in a way that can be experienced by all, and for this reason alone do we accept it as our own experience.

Truth, which causally connects experience with experience, finds an antipode in power, which connects behavior with behavior. One person gives a command and another person obeys because he fears the consequences of noncompliance (Luhmann 1979). Power binds threats with the actions that carry them out. In superficial descriptions of power, it always appears as if only the subordinate person's behavior is bound, as if the dominant person were free in his behavior and could randomly or arbitrarily exercise power. This is certainly not the case. A threat is only credible if, in the case of compliance, the person who made the threat agrees not to carry it out. It must be clear to all participants that only obedience will result in the suspension of punishment. The robber's proclamation, "Your money or your life," only makes sense as an order and threat if the victim believes that he can actually save his life by choosing to forfeit money. Those with power must, therefore, build up the resources to deliver sanctions and also signal that they can suspend such behavior. Modern politics demonstrates the improbability of successfully coordinating behavior with behavior. On the one hand, powerful governments threaten weaker governments with sanctions they do not seem willing or able to deliver, even when it comes to preventing genocide (Rwanda and Sudan) or nuclear proliferation (North Korea and Iran). One might also consider East Germany's (the German Democratic Republic) unwillingness to enforce threats made against its protesting citizens immediately before the fall of the Berlin Wall. On the other hand, powerful nations threaten "rogue nations" with the threat of invasion, even when it seems unlikely that their compliance would prevent sanctions (Afghanistan and Iraq). In such cases we may not observe the medium of power organizing political communication, but rather the noises of propaganda and brute force; which tend to stimulate the social system of the media (Luhmann 1996a).

According to traditional romantic lore, lovers share an emotional bond that drives them to a state of utter dependence. The image of "falling" into a romantic partnership conveys the idea that there was no decision to jump and there is little chance of falling back out. Social system theory does not want to spoil the mystery of love, but it wants to show how love helps partners attribute motivations and make sense of each other's actions. Regardless of what they may feel inside, lovers who want to make meaningful differences upon one another must conform to communication's own "codification of intimacy" (Luhmann 1998). Emotions cannot inform communication, but they can become powerful and compelling themes for conversations in the moonlight. Looking at love as a medium of communication may help us better understand how it can so completely disturb our emotions.

As a success medium, love connects the behavior of one person with the experience of a second person. Every gesture ought to be noticed, and evidence of inattention is far more destructive than any possible change of feelings. At this level of abstraction, connecting behavior and experience, love appears to be a relative of money. However, the giving away of one's money is quietly experienced by all others, whereas the giving away of love is experienced by only one other person. In general, society directs each of us to become an identifiable person and to remain accountable for actions and motives. Each person is given a personal name so that attributions of responsibility can be levied. Love multiplies this to a point near absurdity. Lovers want access to the unmarked depths of total individuality. "Intimacy comes into being," Luhmann wrote, "when more and more domains of personal experience and bodily behaviour become accessible and relevant to another human being and vice versa" (1995a:224). However difficult the task, the one in love expects to find meaning in every aspect of his lover's being (see Fuchs 1999).

Lover's take on full responsibility for understanding and caring for each other's individuality: they each open themselves to the unlimited complexity of the other. Communication—though painfully anonymous—contributes the necessary foundation to every experience of love. Love can and must develop itself against the nameless, public background of society, though every couple confesses love as if they were the first and only ones to have done so. As a symbolically generalized medium for successful communication, the accumulated cultural sediment of love comes to support every couple's desire for intimate understanding. As lovers come and go, the social meaning of love is recorded and made available in romantic stories, poems, songs, paintings, movies, and other semantic resources. Romeos and Juliettes do not have to make up their lines. Red roses, diamond rings, and tables for two are everywhere, and yet there is no loss from overuse. As a medium, love is available to every couple that claims love as its own.

Each of the examples of success media presented above deserves further elaboration. It would be valuable to examine additional success media, such as law (Teubner 1992), art (Luhmann 2000a), and intelligence (Baecker 2002). Our present goal, however, is to introduce success media as coded distinctions that encourage the acceptance of communicated selections for participants in very different functional social systems. Although everything could be done differently, media give the message that participants should behave and experience the world in a limited manner, according to a highly selective form of contingency. One spends or keeps money, pronounces love or keeps quiet, accepts a claim as true or false, activates a threat or tolerates disobedience. The societal systems of the economy, love, science, and politics present dilemmas and wait for participants to determine their own selections. Choosing either side of the binary code can lead to understanding within society, although only one side (the preferred choice) is able to make a connection within the specific functional system. The economy extends itself with decisions to purchase, not to save. Science builds itself by connecting determinations of truth, not falsehood. A social system is reproduced and extended whenever participants choose to accept the preferred side of its distinction (to love, to buy, to obey). A decision for the other side is meaningful and can lead to understanding, but it does not establish a tight coupling within the system that produced the success medium.

As we mentioned above, the underlying logic of success media becomes clear when the scheme of causality is applied to the experience and behavior of different people. We identified this special kind of causality as motivation. With love, a certain experience motivates a confirming behavior; with truth, a certain experience motivates a corresponding experience; with power, a certain behavior motivates a fitting behavior in response; and with money, a certain behavior motivates an appropriate experience. It is important to remember that we are using "causality" with regard to these forms of motivation in a strictly metaphorical sense. We are not implying a physical or chemical relationship between cause and effect. When it comes to making sense of motivation, constructivist psychologists emphasize that observers make *attributions* to order and structure the situation at hand (see Heider 1958; Kelley 1963). Thus, motivation is not really an actual form of causality, but rather an imputed or imagined causality. Success media symbolize that the imputed motivation is meant. The proposed motivation must first be accepted, and by all participants. Then, and only then, is the probability increased that concrete proposals of meaning will be accepted rather than rejected. The resulting successes, for their part, further increase the probability that participants will continue to attribute the same motivations, using them as a resource for connecting future moments of successful communication.

Societal Systems, Success Media, and Functional Differentiation

Our discussion of success media may now articulate a set of theoretical problems related to communication's ability to cause or accomplish different effects that may be anticipated by cultured observers. Which societal problem is solved when the ownership of goods is temporarily fixed by money? Which societal problem is solved when commands are met with obedience because of power? Which societal problem is solved when unbelievable propositions are accepted because of truth? Which societal problem is solved when personally unique experiences of intimacy are socially confirmed as a consequence of love? Social system theory pursues these questions by observing the functional differentiation of society into autonomous societal systems. For instance, to participate in economic communication, buyer and seller must observe the economy as an observer that operates with its own code. The economy prefers that a payment will be made, a tight coupling that will result in connectivity and the extension of its boundaries. Each success medium attaches itself to a societal system, and facilitates a subset of communication that refers to a particular problem. The success medium of money attaches itself to the economy, the success medium of power to politics, the success medium of truth to science, the success medium of love to the assembly of partnerships and families, the success medium of law to the legal system, the success medium of art to the art system, and the success medium of intelligence to the educational system.

From the perspective of the theory, functional differentiation means that the unity of each societal system is achieved by the continued use of its own specifically coded distinction. A variety of societal systems is depicted in figure 5.2, along with corresponding functional distinctions and examples of operating programs. Each of the differentiated binary codes provides a form that enables meaningful distinctions to be drawn about a particular aspect of the world. The political subsystem, and only that system, communicates in the form of politics. Art, and only art, communicates in the form of art. There can be no political communication outside the system of politics, nor is any artistic communication possible outside of the system of art. No matter what distinction a societal system uses for its operations, communication can only happen within the boundaries of society. However, each societal system uses its distinction to observe the difference between itself and its environment. As explained above, the decision to observe or mark one side of the code instead of the other can be described as a "contingent operation," something that the observing system could have done differently. To orient themselves as they make decisions using a functional code, participants in a societal system may refer to programs that identify goals and bundle expected conditions. "A program," Luhmann suggested, "is a complex of conditions for the correctness (and thus the social acceptability) of behavior" (1995a:317).

The binary code of a success medium provides unity within the societal system and replicates itself whenever the system operates successfully. A system cannot communicate anything without recursively referencing its unique code. For instance, the educational system operates on the binary code of pass/fail. Operating according to this code, educational systems select programs that provide the criteria used to inform the difference between passing and failing. Educational programs (religious, vocational, developmental, liberal arts) can change over time, but a system will collapse into uncertainty if its functionally specific code cannot be decided. Teachers cannot pass or fail students without following a selected curriculum, but curricula, educational cultures, and instructional expectations can be changed. A code establishes the possibility of a program, but the program is required to inform or condition the selections of participants. Luhmann maintained, as a matter of fact, that a functional code could not exist without a program that enables the system to recognize a correct selection (2004:34–35). For example, the binary code of politics might be described as the difference between being in and out of power (in power/in opposition). In the United States, members of both the Republican and Democratic parties can distinguish the difference between being in and out of control of Congress or the White House. When the Democrats rule, Congress is expected to pursue a "Democratic" political program (increasing spending on social services, decreasing military spending) and not a "Republican" one. After the next election the opposing party may come into power and Congress may operate with the same political code, but according to a different program.

The thesis of functional differentiation suggests that communication relegates participants to certain channels of society, depending on their intentions. If scarcity is handled as a problem, one must turn to the functional system of the economy—one must use money. Economic problems cannot be solved by writing a letter of complaint to the system's address. If a decision must be made collectively binding in order to have an effect, one must turn to political communication—one must rely on the form of power. Social movements typically set out to "raise consciousness" and "change the way people think." They attain success, however, only after laws are passed and justice is enforced (Luhmann 1996b). If a "discovered" correlation should find its place in the canon of knowledge, if the world's students ought to learn about it—then it must be subjected to the tests of science. In modernity, there is no satisfaction of needs outside of the economy, no generally accepted knowledge outside of science, no effective enforcement of decisions outside of politics, and no romantic feelings outside of love. Functional differentiation means that problems are always solved within the system to which they belong, using the appropriate success medium. Selectivity is necessarily limited to the two sides of an established distinction or

Figure 5.2. Societal systems, functional distinctions, and examples of programs.

Societal System	*Success Medium*	*Functional Distinction or Code*	*Conditioning Programs*
Family	Love	Love/No Love	Blood, Nuclear, Extended, Avuncular
Politics	Power	In Power/In Opposition	Democracy, Monarchy, Political Party
Economy	Property/Money	Payment/No Payment	Budgets, Investment Plans
Legal	Law	Legal/Illegal	Constitutional, Civil, Criminal
Religion	Faith	Transcendent/Immanent	Judaism, Islam, Christianity
Education	Grades, Diploma	Pass/Fail	Curricula: Religious, Developmental, Vocational, Liberal Arts
Science	Truth	True/Not True	Theories, Methods
Protest	Anxiety, Discontent	Protest/Acceptance	Feminism, Environmentalism, Prohibitionism (Themes)

Adapted from Broszіewski 2002a:99–125.

binary scheme, but the indication itself is open or underdetermined. Modern society's combining restriction and freedom, closure and openness, compels the use of success media and stabilizes the reproduction of their forms. Since functionally differentiated systems are by definition autonomous, there can be no central or overarching super code or mechanism of control.

Psychic Systems, Society, and Embodiment

Confronted with the systemic stabilization of success media, individuals may not question whether to put their trust in communication and accept its terms of operation. They have no other choice but to reconstruct the distinctions of society and participate in communication using its success media. Without money, there will be no food. Without truth, no knowledge. Without power, no protection. Without love, nothing but loneliness. One who wants to eat, explain phenomena, live in peace, and escape solitude must at least act as if he trusts symbolically generalized media to help. The codes of success media specify choices in a meaningful manner, making it possible, whatever the choice may be, to make a new connection to a communication chain established in the past. Success media reduce complexity by keeping each train of communication on its own unique, functionally differentiated track.

Success media have developed an effective mechanism that secures their value not only for communication, but also for individuals. This mechanism deploys, in Luhmann's words, *symbiotic mechanisms* (1995a:249; 1997a:378). The term "symbiosis" draws attention to the necessity of making reference within communication to the physical embodiment of participants. As living organisms, participants belong to the biological environment of communication. Symbiotic mechanisms, however, provide communication with ways in which it may allow itself to be influenced or disturbed by the embodiment of participants. Symbiotic mechanisms structurally couple communication and the body without violating the operative closure of each type of system. The differentiation of success media is intricately associated with the differentiation of symbiotic mechanisms. Luhmann argued that:

> Every improbable differentiation of specific function domains must remain moored to the fact that human beings live together in a bodily existence, that they can see, hear, and touch one another... They must have triggering and inhibiting signs ready to hand and plan to form corresponding expectations... It is therefore no accident that all the important function domains must regulate their relationship to the body and that with the differentiation of particular symbolically generalized media of communication this relationship must be tailored more precisely and more specifically to the function by a specific symbolization, namely, symbiotic mechanisms. (1995a:249–250)

While every offer of communication can be rejected or negated, consciousness cannot negate its own body and thus experiences it as a reality that seems independent of both consciousness and communication. The different success media use this inability to negate the body as a way to ensure their own currency and to protect communication against an endless circle of rejection. The success medium of money, for instance, relies on the symbiotic mechanism of bodily needs, on the experience of an observer lacking what the body desires, which can only be satisfied by attaining material goods and services. The success medium of truth relies on perceptions with which an individual may test the consistency of proposed knowledge. The success medium of power depends on the control of physical violence, which must be secured if an individual is to experience the freedom to feel bound or not bound by a decision. The success medium of romantic love relies on the experience of shared sexuality, which, in contrast to language's deceits, appears to directly join the behavior of one person to the experience of another. Due to its immediacy, shared sexuality can register, perhaps more accurately than any other test, when the promised regard for a unique experience of the world begins to fade. Our appreciation of symbiosis does not imply that individual bodies, or the experience of bodies, become parts or elements of communication. Communication uses symbols of experiencing the body and cannot do otherwise. Yet with the help of these symbols, communication gains internal limitations by referencing external conditions. Communication removes itself from control and permits systems in its environment to take over, systems that it cannot itself manage. In order to get what they need, observers volunteer to manage themselves.

A final point leads us back to the fundamental connection between distinction, observation, and communication, as discussed in earlier chapters. All success media of communication are *coded media*. They share this trait with the communication medium of language. This shared quality is exactly what helps them to find their way into spoken and written communication and then, within the system of society, to establish their own boundaries and their own functional operations. Codes, as we described with reference to the yes/no code of language, double the status of all events that can be observed by a system. Anything that can be affirmed can also be denied—this is the law of communication via language. Consequently, a binary code has evolved to order communication within each functional system of society. According to the law of money, anything that can motivate a payment can also motivate not paying. Anything that can push itself through by mobilizing power can also prove to be powerless—and so the code of power doubles the events in the world. What might be true could also be false; and only what could be false might also be true.

"Codes are distinctions," in Luhmann's words, "forms that serve as observational devices. They are mobile structures that are applied differently from situation to situation" (2000a:187). Codes refer to observations, to possibilities for observation, and to observers. During social exchanges, we may quickly and easily determine whether or not someone else is using one of these codes and drawing distinctions between distinctions. We observe a lady at the market standing in front of a shelf, switching her gaze between an item and its price. As we see her observe with reference to the difference between "pay/no pay," we also see that she is not observing the difference between powerful/powerless, true/false, or love/not love. Thus, we draw a distinction between the code that is currently assisting observation, and all other codes, including the code of language. Each code is a specific code that must be distinguished from others. If we fail to make this distinction we slip into confusion, the state of uncertainty that creates the excitement in stories and films about heroes and heroines bewildered by communication. Everyday life must be kept as free as possible from such entertaining confusions, which may be what provokes the impression that modernity is monotonous.

The coding of all symbolically generalized media results in some uncomfortable consequences for individuals. The prospect of successful communication dangles before their eyes. They are motivated. And yet the negative value of each code hangs, like an ax, over each attempt to communicate. Despite the use of skilled labor and quality materials, no one steps up to pay the price for a finished product. Even though a powerful number of threats have been issued, not enough people feel obliged to accept a decision. Notwithstanding immense theoretical and methodological effort, the memory of science must classify the brilliant hypothesis as "untrue." Regardless of the most sensitive, longsuffering observation of a loved one, her gesture must in the end be remembered as a confirmation of unrequited love. For communication, the negative values of the codes are constant reminders of the improbability of all constructions of meaning that promise to achieve and hold on to the positive values. This is how they make learning possible and necessary. One who sets his sights on success must learn to live through failures. A more profound realization is that society offers no all-encompassing form of integration, no metacode that could help people cope with the disappointments and frustrating experiences that result from communication failures.

All of the functional systems operate autonomously. Using their own specific codes, each success medium constructs its own observations and follows its own rules for connecting observations. Through this process of autopoiesis, each functional system produces its own environment and its own world. We can speak about this process and affirm that society unfolds itself in this manner—as we in fact are doing. However, to do this we must use the code of language and, in a book such as this one, we must refer to the code of science. We must say that it is true that success media cannot be integrated. Once something is said and asserted to be true, it can also be contradicted and possibly disproved. One way or another, our observations connect us to a social system, in this case to science. Science potentially lends meaning to our observations, while at the same time limiting our view to its own special vantage point. This also implies: scientific communication tells us nothing at all about what it takes to win success with regard to economics, politics, love, or art. Consequently, individual participants in communication must learn to jump back and forth between codes, without being offered a socially binding set of rules about how and when to jump. Here is the place for participants to demonstrate that they are individuals, whether they like it or not.

If all of the success media can be so easily compared when it comes to their basic structure, why are the related functional systems so different in their operations? Why is there not a single recipe for success? Despite the thesis of functional differentiation, it seems that the observations of economics and politics clearly dominate those of science, education, art, and the others. The theory of success media and their coding offers an innovative approach to these questions. The relative prominence does not result from a hierarchy of functions according to importance. For society, economics is not more important than politics, science, love, or education. For each functional system, its own function remains the sole problem. The theory suggests that the individual codes differentiate themselves according to how quickly and easily they lend themselves to connecting observations with observations. The code of economic communication, pay/no pay, has a strong advantage because it can be so precisely quantified using money, right down to the last penny. The threats with which power builds its sanctioning potential are, by themselves, immediately impressive, especially when it comes to the "shock and awe" of physical violence used by states. In contrast, the code of love is heavily dependent on sensibilities that individuals must develop for themselves and for others. Differences of love must be made to appear within a refined dimension of susceptibility, under emotional conditions that can challenge even the sharpest observers. In a similar vein, the code of science requires a special competence for language and theoretical concepts. Indeed, the difference between true and false can gain traction only within language—it cannot speak to "unmediated" experiences. In different words, some codes are more straightforward than others with regard to coupling and recursively relating observations. For this reason, they muster a higher volume of communication than the relatively more complex codes. One might even argue that the success of some codes is less "improbable" than others; that their effective use requires less conditioning of participants, fewer structural and semantic resources, and lower levels of organized complexity. We may also consider functional systems that have not yet been able to give their code the form of a success medium. For instance, there is the health system with its code of "sick/healthy," and the social welfare system with its code of "help/do not help." These functional systems cannot directly support themselves with the binary problem of acceptance and rejection as handled within communication. Rather, they must refer to people and personal problems. The health and social welfare systems are challenged by the difficulty of socially coding psychological states. The unique characteristics of these codes as resources for observation and motivation help explain the distinctions between the different communication zones of modern society and their relative "weight" or visibility.

As we have already emphasized, success media play an essential role in social system theory. They provide a specific distinction and invite an observer to indicate a choice, promising that either option will lead to a meaningful connection. Symbolic generalizations, in this sense, demonstrate unity in difference. Paying *and* not paying have meaning within the economy. When it comes to love, both loving *and* not loving have meaning. By generalizing, symbolic media organize variety and structure contingency. They close matters in order to open them up. Symbolic generalizations allow each participant in communication to measure up their own selections against an established construction of reality, one independent of themselves and their partners. Whatever their private intentions, psychic systems are restricted to the environment of communication. Only this restriction allows them to use communication to inform themselves and coordinate their own meaning making with the experiences and behaviors of others.

Chapter 6

Differentiating Social Systems

Society, Interactions, and Organizations

Social system theory defines society as an emergent, autopoietic, and self-referential system comprised of all communication. Guided by this unique conception of society, we have so far discussed observation with the use of distinctions that reference meaning, analyzed communication as the basic unit of sociological analysis, described language as the organization of noise with codes, and explored the way society builds differentiated chains of meaning using symbolically generalized media. Now it is time to present a theory of society that differentiates communication into three distinct forms of social systems, each operating within society. These three forms of communication take shape as a consequence of society observing different boundaries—drawing different distinctions—between itself and its environment. Observers of society learn to differentiate these boundaries just as they differentiate the various uses and functions of communication.

From the perspective of social system theory, *society* is only one kind of social system, but it is the one that includes all other possible social systems. *Interactions* are a second type of social system, while *organizations* constitute a third type. The unity of these forms of social systems is established by the fact that each emerges from observed connections between elements of communication (Luhmann 1997a:14). Once again, social system theory takes advantage of a theoretical unity in difference with which it can both generalize and differentiate. A sociological theory would be incomplete if it only addressed one form of social system. Furthermore, a general theory cannot be produced by collecting incongruent theories, even if each one appears to hit its selected target. Sociology gains very little by merely adding up theories of society *and* theories of interaction *and* theories of organization. An "enlightened" sociology must *unify the differences* between social systems; it must become a science of observing differences, relations between differences, and the effects differences have on each other.

It would be fair to reject the premise that society is communication if one could present a better alternative. However, for the past 100 years, as Niklas Luhmann asserted, sociology has made no significant progress toward developing a theory of society: "How can the discipline explain its failure in a task that falls so completely in its own domain and which is so important to its societal reputation?" (Luhmann 1997a:22). It is perhaps an embarrassing deficit, but one that sociologists have apparently come to accept. Consider, for example, how flightily the term "society" is defined in a common reference dictionary of sociology:

> The concept is a commonsense category in which "society" is equivalent to the boundaries of nation-states. While sociologists in practice often operate with this everyday terminology, it is not adequate because societies do not always correspond to political boundaries (as in "Palestinian Society"). Some Marxists, in order to avoid this difficulty, have substituted "social formation" for "society", but in practice these two terms are equivalent. It is more useful to argue that sociology is the analysis of the social,

> which can be treated at any level (for example, dyadic interaction, social groups, large organizations or whole societies). (Abercrombie, Hill, and Turner 2006:395)

In other words, when it comes to the meaning of society, sociologists have nothing much to add to common sense, and what little they do contribute is "inadequate." They use terms that are so undifferentiated that even a tautology ("society" means "the social") promises to be more useful. This is the kind of insufficiency that prompted Niklas Luhmann to argue that sociology needs to progress through its own Age of Enlightenment.

Society as the First Cut

Social system theory points to communication as the social operation par excellence. However, the theory does not exhaust itself in communication theory, nor does it attempt to master everything that could be known about communication. The theory is primarily concerned with describing society, interactions, and organizations as different forms of organizing, restricting, connecting, and differentiating the flow of communication. With its first cut, society draws a distinction, marking off communication "from all noncommunicative events and affairs" (Luhmann 1995a:410). This closure of society creates an environment that permits interaction and organization systems to draw additional distinctions and create additional environments—as long as they continue to work within the boundary of society, replicating the first cut in perfect continence.[1] Regardless of what "commonsense" might say about the boundaries drawn by nation-states, the nested boundaries of society are "independent of natural features like ancestry, mountains, or seas; and as a result of evolution there is only one society: the world society. Society includes all communication and thereby acquires completely unambiguous boundaries" (Luhmann 1995a:410). The theoretical decision to select communication as the encompassing social system results in the following consequence: every connectible communication takes place within the same social system: world society (Luhmann 1997a:150) (see figure 6.1[2]).

FIGURE 6.1. World society as the superset of all communication.

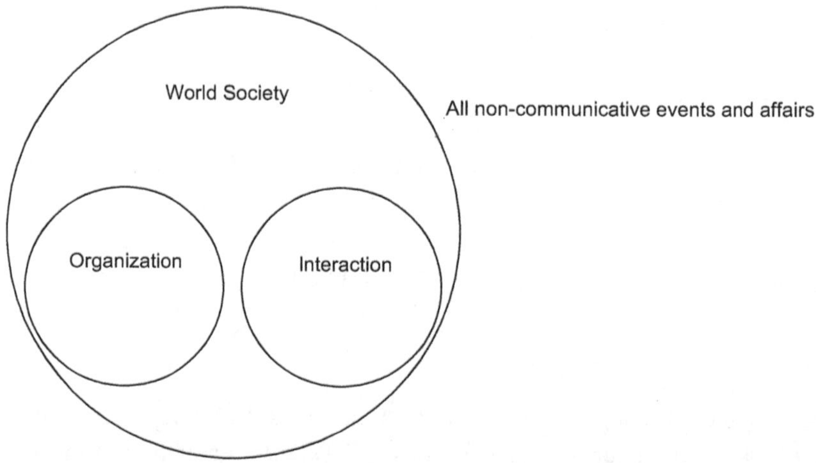

The parallel existence of more than one all-encompassing social system, an obvious possibility, would logically require that the systems share no meanings, connections, or differences (Luhmann 1997a:145). While social system theory has the resources to describe multiple worlds, assuming multiple societies could be observed by the same observer, the tendency is for researchers to nestle multiple interactions and organizations as social systems that operate within the boundaries of a single world society. The rest of this chapter elaborates this differentiation (and unification!) of social systems.

UNITY IN DIFFERENCE: SOCIETY, INTERACTIONS, AND ORGANIZATIONS

Society, interactions, and organizations emerge when different forms of boundaries are drawn within the same material: communication. *World society* marks communicative connectivity as such, as the pure possibility or the widest horizon of communication. Abstracted from any specific information, content, or determination, society supplies every actual communication with reference to other, not-actualized-but-possible communications. It presents potential connections with other information, content, determinations, and so on. Put more simply: while the operational process of communication draws a boundary between what is said and what is not said, society represents the unity of this difference. It "saves" or holds in abeyance everything that is not said within the status of "it also could be said." Only this unity can establish the boundary between communication and everything else. Society can never be present or represented in any single element of communication; it merely provides every communication with the possibility of extending communication. Only by delivering this possibility does society contribute to the meaningful production and understanding of every single communication.

Interaction comprises communication, and thus it can never be realized outside of society. But interaction does not draw the same boundary as society; that is, by distinguishing communication (as a whole) from everything else (as a whole). Interaction draws its own boundary by specifying communication with the distinction of "present" or "absent." Luhmann carefully described the relationship between interaction and society:

> Society and interaction are different kinds of social systems. Society guarantees the meaningfully self-referential closure of communicative events, thus the capacity to begin, end, and form connections between the communications in each interaction. In interaction systems the hydraulics of interpenetration is activated. The push and pull of presence works on those who are present to one another and induces them to subject their freedom to constraints. Therefore society is not possible without interaction, nor interaction without society; but the two types of system do not merge. Instead, they are indispensable for each other in their difference. (Luhmann 1995a: 416–417)

In a first step, the distinction "present/absent" identifies possible addresses for communication, selecting speakers and listeners. Only those persons marked as "present" can speak and listen within the context of the interaction (see Kieserling 1999). Those marked "absent" can neither speak nor listen: they may only be more or less adequately represented by those who are present. Of course, those persons marked "absent" become topics for interactional communication, typically until they come in and join the interaction. Indicating present participants as themes can be precarious and normally requires special conditions. Lovers communicate about each other with enthusiasm, extending their interaction with trust and thereby burdening it with an unusual demand for mutual tolerance and flexibility. In a second step, the present / absent distinction is applied to the objects of communication—to its themes and its other-references. Language provides countless possibilities to present things that are absent from what participants may perceive ("Have you seen my car keys?"). Whatever participants actually choose to talk about appears against this background (the difference between present and absent) as highly selective—and the process of selection itself decides what may be treated as understandable or connectable, and what cannot be. Thus the possibilities and limits of *perception* demarcate the primary environment and referential problem of interactions. Clarifying the distinction between communication and perception, Luhmann observed,

> From the perspective of consciousness, perception frames all communication. Without eyes one cannot read, without ears one cannot hear. Communication must be highly conspicuous in the perceptional field if it is to be perceived at all. It must captivate perception—by means of some striking noise, through bodily postures explicable only as expressive behavior, or by employing special conventional signs in writing. (2000a:14–15)

Consciousness, in contrast to perception, is the primary feature in the environment of society or communicative connectivity as such; and, thus, questions about perception are rarely entertained. Luhmann suggested that consciousness "compensates" for the operative closure of neurologically based perception, in the same manner that the social system compensates for the closure of consciousness (2000a:10). While consciousness uniformly absorbs sensations independently originating in the nose, eyes, ears, and fingertips; the social system absorbs the independent consciousnesses of its participants. At the same time, the mechanism of perceiving that others perceive differently (the perception of perception), provides a major generator for communication and socialization in general. Interactions confront consciousness with the recurrent problem of choosing between presence and absence. Operating with this distinction, interactions provoke consciousness to activate perceptions that can test and contest what communication has achieved so far and what it might realize in the future. This kind of activation guarantees that the perception of perception and of being perceived remains stimulated and will somehow relate itself to what is possible as communication and what is not. As an illustration, consider the normal assumptions an instructor makes while lecturing to a group of students: the instructor can reach students by making perceivable sounds (lecturing in a common language) and introducing visual aides (writing and drawing on a chalkboard, using an overhead projector, indicating objects). The instructor's assumptions about what works within a classroom interaction change, however, when a deaf or blind student is present, when students cannot understand his language, or when students choose to exclude the instructor from their ongoing interaction with fellow students. In such a manner, communication sorts out the assumptions of participants.

Organizations, the third type of social system, produce a different bounded unity by identifying elements of communication as decisions (Luhmann 2000b). Compared with society and interaction, where one can distinguish between "communication / non-communication" and "present / absent" respectively, it is more difficult to indicate the other side of decisions. This difficulty of describing the unmarked state is a basic characteristic of organizations. Consequently, we must devote more time to describing the operative base of this type of social system. We cannot refer simply to decision / non-decision, because this would not clarify the boundaries between different kinds of organizations. It would also not support differentiation between different kinds of decisions, many of which may not be attributed to organizations (the decision to marry, for example). To assess its potential for selectivity and exclusivity, that which establishes the organization's own kind of autopoiesis, we have to clarify the notion of decision used by the theory of social systems.

There are no decisions without alternatives and without the possibility to mark a preference for one of the alternatives. This structure reveals that decisions are a special form of observation, of distinguishing and indicating both a selection and alternatives. To make a decision entails cutting, as does the drawing of every distinction. We may appreciate this by considering the origin of the word in the Latin term *decidere*: to cut off. An observer's alternatives must be distinguished from everything else and from each other, and only one of them must be indicated as preferred. The decision process results in the indication of all not-preferred alternatives as *excluded*. This hints at the unique quality decisions have when compared to normal observations. Normal observations indicate one side, organize connections to that side, and then forget the unmarked side of their distinctions as soon as the connection is realized. In contrast, decisions indicate and memorize that they exclude something: *decisions indicate both sides of their distinctions*. If an organization forgot what its decision excluded, the decision would no longer be a decision; it would regress to being a simple fact. In whatever forms decisions appear in everyday life, in historical events and in the day-to-day operations of organizations, decisions serve communication and consciousness as the memory of contingency. Decisions proclaim: things *could* be different, although they are always just as they are.

Decisions always involve uncertainty, whether it appears more on the side of the facts (the descriptions of alternatives) or on the side of the values involved in marking a preference. To cope with this necessary uncertainty, decisions are typically supported by appeals to cultural resources that lay outside the realm of the actual decision itself. For instance, decisions may be supported by reference to what can be taken for granted, common sense, customary practices, and allegedly shared values or programs. In this mode, decisions are not differentiated against communication in general, and there is no apparent differentiation of decision-making systems. Every organization does not need to reiterate the difference between male and female restrooms, between domestic and international postage rates, and between workdays and holidays.

In a second mode, decisions refer to the socially memorized results of previous decisions. A decision may be identified as one moment within a network of decisions, with each connected decision relaying successive achievements in the reduction of uncertainty. Consider the following network of decisions as an illustration of this sort of relay: imagine that somebody once decided that you have what it takes to make a certain decision (and not another), and so you (and nobody else) better decide what should be done (and what should not) and when (now and not later) you will announce your decision (and not something else) so that others can inform their own choices. This is the autopoietic mode for making and connecting decisions within formal organizations, a mode that appeared very late in the socio-cultural evolution of society. Perhaps the earliest precursors of modern organizations were military and religious organizations that constructed hierarchal decision-making chains, specialized roles, and structures to memorize and recursively relate past consecutive operations. In contrast to interaction and society, the organization is not a historically universal form of social system. Rather, its appearance is a significant feature of modern, functionally differentiated society (Luhmann 1997a:827).

Organizations decide which persons may become members and which may not. They decide what tasks members will tackle and with what kinds of programs. They decide which communication channels will carry the open space required to process new decisions. Within the nexus of these three distinct kinds of decisions (members, programs, and channels), organizations select which decisions do and do not concern them. The memory of exclusion becomes a decisive moment in the production of every single organizational decision, in sharp contrast to the common-sense orientation of normal decisions as mentioned above.

Now we can specify the operative distinction that contributes closure to communication within organizations. This form of system is bounded by the distinction: "decision-within-a-network / not-a-decision-within-that-network." The increasingly complex identity of a network reenters every subsequent decision within the form of the organization. In order to support this elaborate and burdensome autopoiesis of decision making, organizations produce self-descriptions, ideologies, texts, records, rules, culture, and rituals. These support structures help the organization reproduce its identity, decision after decision. According to Karl Weick, there are identifiable processes that create, maintain, and dissolve social collectivities, and the ways in which these processes are continuously executed *are* the organization: "Thus if you wanted to learn something about an organization, you might look for interlocked behaviors that are embedded in conditionally related processes" (Weick 1969:1–2). The unity of the organization is self-produced by the organization's own decision-making behavior. In Luhmann's formulation,

> Since [organizational] memberships are founded in decisions and the subsequent behavior of members in decision-making situations is dependent on the membership, one may characterize organizations as autopoietic systems that operate on the basis of communicating decisions. They produce decisions out of decisions and are in this sense operationally closed systems. In the form of the decision there is also a moment of structural uncertainty. And because each decision evokes other decisions, every decision reproduces this uncertainty. (Luhmann 1997a:830)

Once again, we note the importance of observing differences: social system theory distinguishes between society, interactions, and organizations. Furthermore, it assumes that these three types distinguish themselves by using specific distinctions to achieve operational closure. Each type of social system has its own binary code, an operating form that cannot be used by the other system types (see figure 6.2).

Sociologists have already developed important theoretical approaches to understanding the three types of social systems. However, they have done so in a fragmented, territorial manner, and they tend to overgeneralize their findings. We can attribute, for example, functionalism and structuralism to the observation of society, symbolic interactionism to the observation of interaction, and some recent approaches of institutionalism and network-analysis to the observation of organizations (or at least,

FIGURE 6.2. Types of social systems and specific distinctions.

Social System	*Specific Distinction*
World society	Communication/non-communication
Interaction	Present/absent
Organization	Decision-within-a-network/not-a-decision-within-that-network

to forms and processes of decision-making networks). A plethora of data about the special types and features of social systems has accumulated under the guidance of these divergent sociological theories, producing a body of knowledge that should not be rejected, repeated, or replaced by social system theory. The theory does not aim to prove that it has something to add to every single other scientific description. Sociology has drawn and maintained boundaries for theorizing about the different forms of social systems, and this reveals important insights about the discipline's intellectual heritage. But whatever might or might not be achieved in this direction, the theory asserts that none of the specialized theories can adequately function as a base for a general theory of social systems. A general social theory does not emerge from generalizing the assumptions and findings of theories of societies, or of theories of interactions, or of theories of organizations and decisions. A general theory must start with differences and search for a meaningful way to construct unity. As social system theory suggests, a general theory might start with the difference between system and environment and, after indicating the observing system, proceed to internally differentiate forms of social systems.

We shall now elaborate some of the potential insights of a sociological analysis of society that begins with differences. Specifically, we will explore the differences between society and interaction, interaction and organization, and society and organization as determined by the unique binary form that closes (and identifies) the communicative operations of each system.

Society and Interaction

The best description of the boundary between society and interaction is offered by the term "episode" (Luhmann 1995a:406). It makes clear that the difference itself (society/interaction) affects both sides and can never be reduced to either one of the two systems. Episodes mark events with a beginning and end that mutually identify each other. An episode is the unity of the difference between a beginning and an end. On one hand, no interaction could be started if there were not a possibility for it to end. The beginning of a marriage is celebrated, not devastated, by the utterance, "Until death do us part." On the other hand, episodes may only start and end within the constant flow of communication. As Luhmann described it, "The beginning and end of an interaction are merely caesuras in society's autopoiesis. They serve to achieve structures that cannot be made congruent with society and yet equip it with complexity by building in differences" (1995a:407).

What kind of differences do episodes supply to society? Interactional episodes add complexity to communication by referencing and connecting the perceptions and concerns of present participants. Interactions rely on society to provide understandable beginnings and endings, but at the same time offer the external (environmental) stimulation that only human participants can make by their presence and active participation in communication. Communication, Luhmann reiterated, is a closed system that cannot touch the environment that sustains it:

> Just as an organism does not live outside its own skin, or a psychic system operatively extend its consciousness into the world, or an eye create neural contact with what it sees, so a society cannot communicate with its environment. It is completely and without exception a closed system. This distinguishes it from all other social systems, in particular from interaction systems, which include communicative relations within their environment, welcome what is new, utter decisions, and so on…
>
> The closure of recursive communicative relationships does not liberate the system from the environment. It is and remains dependent on sensors that convey environment. These sensors are

> human beings in the full sense of their interpenetration: as psychic and as bodily systems...[Interpenetration] enables the emergence of autopoietic systems by opening up environmental contacts on other levels of reality. (1995a:409–410)

This feature of interaction constitutes a very important source of variation for society, and thus stimulates social evolution. Deviations, innovations, surprises, and misunderstandings that occur on the level of interaction may be selected and carried beyond the single episode, occasionally impacting future episodes as expectations.

Before the advent of reading and writing, meaning could be produced and reproduced only within interactions (including its designated signs and symbolic objects). "Under primitive conditions," Luhmann maintained, "one would assume a societal reality that is very like interaction, that does not yet have the effect of giving form to interactions but is constantly revised as they take place...[All social forms] remain bound to concrete localization, and must be present in order to have an effect" (1995a:417). Between successive episodes of interaction, society is present as a reference to all other episodes available in social memory, by allegations that certain connections between episodes must be respected, and by narratives about grand episodes that hold everything together. Ritual is the kind of grand episode that represents this form of society within society: the society of interactions performs itself as interaction—offering no social outside, no vista for mere observers (such as reporters, tourists, scientists), and no meaning to critics. Deviation—the starting point for innovation and evolution—is dangerous and unlikely. Since there are no documents with which to compare old and new forms, any departure from patterns carried by social memory is both difficult to effect and recognize. This lack of innovation must not lead us to overestimate the stability or underestimate the flexibility of oral cultures (see Goody and Watt 1963). The point is that forms are used and occasionally modified, and that changes are not registered as change because they are not described, recorded, or formally controlled. The forgetting and remembering of forms is governed, as Jack Goody and Ian Watt suggested (1963:30–31), by the "homeostatic organization of the cultural tradition." It took the arrival of bureaucrats and anthropologists, with their techniques of writing and comparing, to characterize the "traditionality" of oral cultures in contrast to the means of social forgetting and remembering of literate cultures.

Writing and printing—over a period of several thousand years—led to an increasing gap between interaction and society. This differentiation enabled communication to increase its capacity for abstraction and achieve greater independence from interaction (Luhmann 1995a:422–426). What cannot be said immediately can be written down and disseminated at a later time. What cannot be observed in co-presence can be studied and evaluated through descriptions on another occasion. This does not mean that interactions became unimportant, but they did lose their dominant and exclusive exercise of social control. With communication in writing and print, society could be experienced as the permanent social absence, the nowhere of social reality. For the first time, society could confront individuals with the impersonal "They" ("*das Man*") (Heidegger 1996:127) and the "generalized other" (Mead 1934:154); a confrontation resulting in the written semantics of "alienation." As a theory of differences, social system theory points to the fact that both sides of the distinction have become more complex because of the growing repertoire of episodic forms and of available kinds of interactions. What we no longer have is one distinguished form of interaction that could represent society, in contrast to former societies with their "great rituals." Even ceremonies involving kings, popes, and presidents can no longer be more than spectacular events.

One major consequence of the differentiation of interaction and society and the increase of complexity on both sides is that the body is decoupled from any kind of immediate social meaning and order. On the one hand, the living organism provides consciousness with sensations, with a base for perception and for the perception of being perceived. Society, on the other hand, through socialization, language, and culture, provides suggestions for how to form and organize this raw material as if it had significance. Interactions require the processing of such forms with the distinction present/absent, selecting some forms as present while holding as absent everything that could otherwise have been possible. The bodies of participants in interaction must be integrated into this double game of form and presence. There is no super-form for this task of integration. In the course of socio-cultural evolution, the body was set free to become both a medium of conscious identity and a medium of society, as we described above with reference to the symbiotic mechanisms of success media. Every single interaction constitutes a *temporary* coupling of body-forms within both media. Every form dissolves with the end of every episode, and new ones will be established by new beginnings. In modern society, ritual can be recognized as the remembering and repeating of valuable forms of mutually coordinated body-behavior. We can decide to participate in ritual, as long as we know the rules (Lee 2005a). The rituals of traditional societies lacked this moment of decision and uncertainty (see Ong 1982). There was no reflected gap between interaction and society, thus the body (the means of both perception and presence) was always presenting and representing all that was.

Interaction and Organization

In order to describe the consequences of the difference between interaction and organization, we can draw on the concept of the role. In sociology, a role is usually described as a set of behavioral expectations or "norms" that are anticipated from a person who occupies a distinct social status (Linton 1936; Merton 1957). An actor's performance in a role may result in observers applying positive or negative sanctions (rewards or punishments). There are roles that are not directly related to organizational positions: the role of the mother, the seducer, the roles of the gentleman and lady, the sick one, the guest and the host, the stranger, and so on. But the overwhelming majority of possible social roles may be clearly attributed to organizations. Operating as "decision making machines" (Nassehi 2005), organizations rationalize and crystallize connections between recognized social positions and expected forms of behavior. Within the context of an interaction, observers may refer to the same organizational decisions to assess, manage, and coordinate one another's performances within established roles (Lee 2005b).

All in all, we can take the emergence of clear-cut roles as an effect of modern organization. In a socio-evolutionary perspective, formal organizations and rule-oriented behavior introduced the sharp distinction between person and role (see Luhmann 1995a:316; Turner 1978). This distinction did not exist before and must still be denied whenever the relevance of an organization must be postponed or discontinued, as when a soldier comes home to a family, when a teacher falls in love with a student, or when a judge is offered a bribe (see Luhmann 1997a:771–772). Each person participating in an interaction can always also act in another way, Luhmann observed, and may or may not fulfill officially designated roles, rules, and expectations—*but not as a member of an organization* (1997a:829). Organizations, with their decisions, organize and limit the variety of meaningful, connectable behaviors.

Filtering complexity by means of membership and its complementary roles (client, customer, supervisor, consultant, etc.), organizations provide interactions with indifference toward personal specifics and needs. Roles abstract from individual and biographical particularities. Interaction typically requires signs of recognition, if not a reciprocal show of respect and appreciation for those present. Drawing formal distinctions between the person and his or her role liberates interactions from carrying this cumbersome burden. Organizations generally meet individual needs for self-esteem with a blind eye, compelling restraint on the presentation of the self in organizational life. When it comes to observations made by an organization, job titles and personnel numbers are more useful than the personal names or characteristics of participants. Formal networks encourage this indifference to the personal in order to establish and enforce otherwise improbable foci of interactions; such as the selection of the organization's means and goals, and adherence to rules of established procedure. Max Weber (1972:650–678) was one of the first to draw attention to the fact that bureaucratic observations gain special meaning because they are impersonal. The rigid and impersonal qualities of interactions within organizations may be lamentable, but what appears to be a cold insensitivity merely functions as a principle to regulate the boundaries of two types of systems. It serves to couple interactions and organizations, thereby enhancing the possible complexity of both. Organizations can make far-reaching decisions without having to observe the biographical details and personal tastes of those members involved in their implementation and consequences.

We want to illustrate our very general remarks about the boundary of interactions and organizations with the example of instruction in elementary school—a highly improbable kind of interaction and mutual perception. Instruction happens as interaction between a teacher and students. To appear as a pupil in a class, an individual must first be separated from his or her family (from parents, siblings, pets, favorite toys, and so on). This separation would not work without decisions: where the kids should go to school, what time they should arrive, who will teach them, etc. As Robert Dreeben elaborated in an insightful study of schooling (1968), it is mainly the separation from family that socializes a young student into the more "formal" modes of organized society. The point is that the pupil will inform himself about how to cope with class situations by comparing and contrasting experiences from his family life—thereby learning the difference between person and role. The main problem of elementary teachers is not to introduce useful knowledge, but to support and enforce this kind of social differentiation. In other words: elementary teachers have to support the self-transformation of children into pupils: instructing the difference between interaction and organization.

From the home and family to the school and classroom interactions, children learn to cope with organizational indifference to persons in a twofold manner. First, they are together with other children of the same age—not with brothers and sisters of different ages. In the family, an older child may assume she will be treated differently than a young sibling: staying up later, enjoying more responsibility, having more freedom from parental control. In the classroom, sibling order is a variable that cannot be considered, along with seating preference, favorite colors, shyness, bladder control, and other details. At home, and in the family circle, children have to account for their personal uniqueness and the memory of all concrete interactions. But at school they have to look for new sources of information to account for differences in treatment by adults. Pupils learn that in the organization of school, performance and the memory of past performance is what counts when it comes to the distribution of grades and other positive and negative sanctions. The acceptance of this rationale is marked by self-identification as a pupil (that might lead to the conscious decision to choose a path of deviance).

As they advance from grade to grade, students learn that their teachers change, but that each performs the same role within interactions. Year after year, classroom after classroom, the teacher is always that singularly marked person in front of the group of pupils, distributing the right to speak and commanding the focus of attention. Taking both features of classroom interaction together, the children learn to handle indifference toward their own person, along with indifference toward the person of particular teachers. Each pupil may recall that they have parents who love them and who seem interested in the fullness of their inner complexity, but each must learn that parents are excluded from the classroom interaction. Learning which interactions may and may not inform other interactions is itself one of the most critical elements of the informal curriculum.

It goes without saying that this indifference for persons is not an unlimited one. If a child suddenly begins to cry, the teacher and other children cannot ignore the child's unique constitution and simply go on with the programmed class interaction. Nonetheless, there is no general meta-rule for creating and modifying such limits. Neither interaction systems nor organizational systems function as machines do. Boundary determination remains a problem and a source for creativity and negotiation. The variability of the limits of indifference to persons constitutes a huge field for empirical research. In the context of our example, the handling of this variability also constitutes a practical challenge for those role-takers who are expected to control and manage the unfolding of interactions within organizational settings. Within a long tradition of sociology and pedagogy, teaching is regarded as a profession. One of the major characteristics of professionals is that they have to settle the uncertainties for which society (and its organizations) cannot provide a general solution. Inside the classroom, the management of students/children by a teacher/adult belongs to this sort of uncertainty. Guided by the necessarily vague and imperfect limits of professionalism, certain role players within organizations are formally selected to make and be held accountable for their own decisions.

Society and Organization

As mentioned earlier, the difference between society and organization belongs to the description of modern society. Both are autopoietic systems, referring to and differentiated within the same material: communication. They presuppose forms of consciousness in their environment—individual psychic systems that must be able to separate themselves as role players from their immediate perceptions and imaginations, and be capable of constructing identities that outlast the short-term communications of interactions. The mediating form is the basic role of membership. Membership is marked—in sharp contrast to pre-modern participation—as a revocable decision, as a contingency. The membership decision is always a two-sided one. It hints to the individual (and its internal complexity) as well as to the organization (and its corresponding complexity). In this form, membership can be made conditional, represented especially in the form of privileges and duties that are subject to other decisions (excommunication, cancellation, discharge, release). The mark of membership—like the mark of communication for society and the mark of presence in interactions—marches through the realm of all organizational communication. It establishes and normalizes (routinizes) expectations that would be improbable and impossible without this special achievement of socio-cultural evolution. At this point we concentrate on two major examples of such probable improbabilities: the form of the *career*, and the form of the *program*. Both phenomena have to be conceived as boundary terms, gaining their identities only by referring simultaneously to both sides of the distinction between society and organization. With regard to history, our thesis is that neither the career nor the form of functional programs could be found in older societies that could not distinguish organizational communication.

The career has become the "most important mechanism of integration" between individuals and modern society—surpassing older forms like "clan solidarity," "one's family origin," or even "moral solidarity" (Luhmann 1997a:742). The career is conceived as a step-by-step process, laid out along the lifetime of individuals, in which one step is dependent on former steps and each progression limits future steps. Every single step—this is the significant feature of the career—combines some kind of self-selection by the individual with a complementary selection by other individuals or third parties. In short, the career relates and orders sequences of self-selections and hetero-selections. Positions in organizations are the most visible and quantitatively most important elements of the career. Without membership in at least one organization, a modern person could hardly survive socially and materially. Self-selection means that the individual decides to exclude many other things that he or she could do in order to fulfill the demands of the position. Hetero-selection means that the person is selected and constantly evaluated against a concrete or abstract background of other possible candidates for the same position, and that the description of the position may be adapted to the changing decisions of the network.

Along the boundaries between organizations and society, the form of the career modulates nearly every social address. It determines who the participants are, what kind of demands they are able to formulate, what kind of duties they have to fulfill, and so on. As a boundary marker, the career is not subject to organizational control. It is society that limits what could be generally expected in certain positions and in certain stages of careers. Organizations merely use these limitations to produce their own combinations. Furthermore, there are other mechanisms of using the form of career, like networks of personal contacts, professional reputation, and resonance within mass media and public communication. But organizations monopolize the field of normal careers, and everything else has to be established in difference to that normality.

The second improbability that is established at the boundary between society and organizations can be described with the term "function-oriented programs." Society is differentiated into its functional subsystems, which can be recognized by the exclusive use of their specific binary codes. As described in a previous chapter, paying/not-paying identifies the operations of the economy. Government/opposition guides the operations of the political system. True/untrue separates science from common communication about facts and opinions. The problem of these codes is that they do not provide information about which of their sides should be chosen in a concrete case. It is exactly this deficit that is compensated for by specific programs. Programs contribute supplements that make it possible to decide between binary alternatives. They discriminate between right and wrong—but each program does so only within its operational field. There are economical programs, such as the investment programs of enterprises, budgets of public administrations, or the priorities of private households. There are political programs, such as legislative texts for the executive level or the programmatic texts of political parties. There are scientific programs, such as theories and methods. There are also programs for art, certain styles or classical paradigms. Programs such as these prescribe when to pay, when to engage in public conflict, when to test or contest a new hypothesis, when to accept or reject the sensations caused by a work of art, and other matters. In this context, functional differentiation precludes learning from a decision in one dimension about how to make a decision in another dimension. A beautiful painting may not necessarily be an economic success. Receiving a doctoral degree may not make one a successful father. A criminal record for drunk driving may not infringe upon one's right to participate in an election. The old-world notion of the unity of the good, the true, and the beautiful has passed away.

All programs are developed, discussed, and criticized on the level of society and related to the functionally specific codes. If this were not the case, functional differentiation could not have been accomplished. But society cannot decide programs! Decisions exclude alternatives, and society preserves the excluded possibilities and thereby the contingency of all decisions. Only persons and organizations can articulate, select, and decide to implement programs. Most programs are related to personal decisions: what to buy, who to elect, what art works to display, who to love, which truths to accept, what plan of salvation to adopt, and so on. But as individuals demonstrate idiosyncratic tastes and changing concerns, persons are not a reliable source for long-term stability and shared expectations.[3] Thus, important and complex programs have to be implemented through the form of the organization. Organizations formulate goals, orient many peoples' actions towards these goals, and force every change into the narrow and sluggish channels of organizational change (see Brunsson and Olsen 1993). In the form of organizational guidelines, functional programs gain stability without being completely exempted from criticism and possible variation. You don't have to think about what car you will buy when yours no longer works. You don't know how cars will look in five or ten years. General Motors, Ford, Volkswagen, Toyota, and other organizations are making sure that your desire for mobility will find an appropriate solution. Didactical and pedagogical concepts may be debated endlessly in academic circles and in public communication. However, schools will somehow adhere to their day-by-day goals of educational success, and to the goal of fulfilling parental expectations about the right kind of education. Without such "trust in systems" (Luhmann 1979), no preference for this or that school could justify the decision to release one's child to external observers for many hours, days, years, and possibly decades of education. The stabilization of programs through the organizations that observe them constitutes at least one means, at least for the time being, of coping with the uncertainties of right or wrong.

INSIDE/OUTSIDE: A PRIMARY DISTINCTION OF SOCIETY

We have asserted that social system theory approaches society by drawing a primary distinction that includes society on the inside, and everything that is not communication on the outside. On the inside of society, social system theory observes the subsequent differentiation of communication into world society, interaction, and organization. When it comes to each of these types of systems, the primary distinction between inside and outside is recursively indicated. The inside is described as the social system, while the outside is described as the environment.

With respect to society, sociology has traditionally operated with a different primary distinction, one separating the whole and its parts. In this manner, interactions and organizations are considered parts of society, which may also include all kinds of other parts. In fact, beginning with the whole/parts distinction, it is difficult to determine what is not part of society. The inside/outside (system/environment) distinction is preferred because it provides a boundary that permits the meaningful observation of difference, as we have argued above.

Social system theory is not concerned with decomposing a whole into parts or divisions and then explaining how society is more or less than the sum of its parts. Dependence on this familiar strategy is to blame for sociologists having to rely on common sense to define what they mean by society. Sociologists cannot impress anyone with the claim, "We can't identify society either, but we know it's there because we can see its parts." The differentiation of society implies that each social system reconstitutes the world society to which it belongs and which it reproduces through its own operations, through its own difference of system and environment (Luhmann 1997a:598). "If one begins with communication as the elementary operation whose reproduction constitutes society," Luhman argued, "then it becomes apparent that world society is implicated in *every* communication; and this remains the case regardless of the concrete subject matter and the spatial distance between the participants" (1997a:150). Through its own continued process of differentiation, society multiplies itself *inside* itself, constructing new distinctions between systems and environments within communication. This increasing complexity is not centrally guided by society or in need of overarching coordination or integration. Always operating inside communication, each social system establishes its own inside as it reproduces its own boundary. Social systems good and bad, moral and immoral, small and large, central and peripheral, real and virtual draw their own boundaries within a single, but polycontextural world society. If the primary distinction of inside/outside is clearly drawn, then it can "re-enter" and inform subsequent distinctions of communication, supporting the observation of differences within society, interaction, and organization. In Luhmann's formulation,

> System differentiation is then nothing more than the recursive building of a system, the application of system building upon its

> own result. In this way, the system in which additional systems are produced is reconstructed through an additional distinction between subsystem and environment. From the vantage point of the subsystem, the rest of the encompassing system is now environment. Thus the whole system appears for the subsystem as a unity in the difference of subsystem and subsystem environment. System differentiation generates, in other words, environments internal to the system. (1997a:597)

When it comes to society, we repeat, social system theory starts by drawing a distinction between communication, on the inside of the system, and everything else in the outside environment.

A Sociological Distinction

With our discussion of the types of social systems and the consequences of their differences, we have only introduced some of the highlights of the theory of society. Social system theory actually offers much more of an opportunity to further discoveries than a system of canonical knowledge does. The theory offers a valuable arsenal of additional concepts that could be fruitfully applied in sociological research. The value of using the concepts of medium and form, for instance, could be tested in countless empirical contexts. We could critically evaluate the notions of structural coupling, semantics, self-description, operational closure, observation, and world society; checking each to see what it might contribute to our images and knowledge about society, interactions, organizations, and the differentiation of communication. We could reintroduce the difference between cooperation and conflict that was so central for classical approaches to social theory. We could challenge the thesis that society, interactions, and organizations are the only forms of social autopoiesis and system building. We could test the social status of collective movements, of communities, of groups, of subcultures, and of congregations. In this chapter we could not provide more than a hint of the potential rewards of thinking in terms of differences. Sociology, armed at last with an adequate theory of its fundamental subject, can finally develop its potential for observing communication and making meaning of society and its different social systems.

At the end of this chapter, it makes sense to consider the status of the sociological observer. It should be evident that sociological communication can only make its connections inside society. Every sociologist is involved in communication before he or she decides to go to a particular research site and focus his or her attention on identified objects or subjects. Field researchers must read many books and articles before their own writing can inform others, and "arm-chair theorists" begin to theorize only after studying "real" texts. Gaining empirically based knowledge about this or that interaction or this or that organization requires involvement in what is or might be outside of oneself—what is excluded by oneself—and therefore requires an informed imagination. The sociological observer is the one who decides to be organized and informed by the sociological imagination (Mills 2000), the historically assembled self-reference of the discipline, while also marking the boundary between sociology and its external reference. Oriented by the social memory of the discipline, the sociologist imagines that redundancies may be observed and documented in the field, despite the incessant variety of social phenomena. The sociologically cultivated observer expects to see the same things appear and reappear in different settings, even while hoping for surprises that might compel the accommodation of self-referential expectations to the "factual conditions" supplied by external reference (Lee 2007:461).

Sociology is only one of many possible horizons of meaning from which to construct identities and differences, but its boundaries must be wide enough to describe all types of social systems, including sociology itself. Interactions and organizations, for instance, can be described sociologically because sociology has prepared observers to recognize its own familiar unities in difference. Sociologists operate by observing the conditions under which first-order observers and second-order observers reference any kind of social system to inform and coordinate their own observations. Social system theory offers a general theory that makes sociology's internal unity and its openness to differentiation possible.

ENDNOTES

1. "Distinction," according to George Spencer Brown, "is perfect continence" (1979:1).
2. Figure 6.1 has didactic value, but the visualization unfortunately suggests space and spatial relations. It also appears that society is a container for organizations and interactions. Readers may prefer to think of the circles as units with boundaries that communication must continually compass.
3. Doubting the idiosyncratic nature of consciousness, Peter Fuchs provocatively asserted that, against all traditional assumptions, consciousness is neither idiosyncratic nor private. Even when observing itself, consciousness makes use of the generalized medium of signs learned through participation in society (Fuchs 2004:125). Nonetheless, organizations depend on programs to manage the use of signs by their members.

Chapter 7

Forms of Society

Connectivity and Boundaries

How does one define society? Does it exist? Does it have an address? Is society good or bad, healthy or sick? Can you get kicked out of society? What parts of society change and what parts always stay the same? Was there or will there ever be a Golden Age? Are some societies better than others? What is the best form of society?

In our everyday conversations about society, we entertain ourselves with all kinds of interesting questions without ever articulating exactly what the term signifies. Each person seems to understand the meaning of society—its forms, nature, and essential characteristics—in a slightly different way; and yet, we seem to be able to understand what we are talking about. Perhaps we tolerate the vagueness of the term because we typically value having good conversations more than having conversations using well-defined terms. After all, we are not always willing to work hard enough to please the scientists and philosophers. Sometimes we just want to talk.

Though loose and casual discussions about society can be amusing, the usual fuzziness of the concept stands in the way of a disciplined attempt to gain, transmit, and consolidate knowledge about what can be indicated with the word *society*. Put in a different manner, as long as each of us has our own personal definition and image of society, it seems unlikely that we will ever arrive at any shared understanding or expectation. The absence of any common perspective on society leads to an option that should be considered very seriously: we might just exclude society as a possible object of scientific knowledge. The concept of society may be so abstract, subjective, or ideological that there may never be a chance to qualify the term with a specific indication. Without first clearly distinguishing what we mean and do not mean with the term, we cannot possibly make society a subject of scientific discourse.

The option of excluding society from scientific inquiry is not as absurd as it might first appear. Indeed, some sociologists have already considered the possibility. Friedrich Tenbruck (1981), for example, argued that the term was so overly steeped in values and interwoven into ideological debates that it could never guide sociological theory and research. As an alternative, sociology ought to pursue a science of "the social."

In a more implicit manner, sociologists have essentially rejected a science of society by failing to create a specific notion of society that could be distinguished from its use in any non-scientific context. When sociologists discuss society, they tend to demonstrate more confusion and disagreement than non-sociologists. They spend their time dazzling readers with provocative observations of what other sociologists have called "society." Sociological theory, in particular, seems to have committed itself to theorizing about men and women who write sociology. Nonetheless, as Luhmann concluded, no genuine progress has ever been made toward constructing "even a partially adequate *scientific* theory of society" (1997a:17). After all, though sociology cannot do without science, science made tremendous progress prior to the appearance of sociology. Instead of helping us understand our "hypercomplex" social system, our scattered, worn-out methods of communicating as sociologists lead us to produce research that is fragmented, trivial, moralizing, and intellectually medieval. Perhaps it is time for sociologists to give up pretending that they know anymore about society than anyone else.

Sociologists must face a hard fact: to produce knowledge about society we must make a rather painful decision. Making this decision constitutes a deep cut into our habits of thinking and arguing. When it turns its gaze upon society, science cannot escape the problem of guaranteeing its own possibility—it must establish conditions for its own functioning. In order to produce "scientific" statements, the central terms of these statements must represent clear distinctions. The term "metabolism," for instance, indicates certain biological processes and leaves out other biological processes that have their own specific terms. We may begin to understand the difference metabolism makes only after it means something by not meaning everything. In the language of social system theory, we make meaning for ourselves by closing down wide possibilities in order to open up narrow ones. Science's own distinction—the difference between true and false—results from a decision not to consider other distinctions: for example, legal/illegal, sacred/profane, good/bad, profit/loss, new/old, exciting//boring, etc. By narrowing its function to a specific difference and code—by excluding all other options—science opens itself for solving a wide variety of problems. In order to establish society as a possible object of scientific inquiry, the term "society" must also be marked by a clear distinction that can inform the difference between true and false. This—and only this requirement—leads us to the painful decision announced above. For the future of sociology as a science, social system theorists declare that *society does not consist of human beings.*

This assertion does not in any sense imply that human beings are unimportant or that they must be neglected by the science of society! Such an implication would miss the aforementioned problem: to establish a possible object of knowledge. The conceptual unit commonly called "human being" is at the same time too much and too little to be the reference point for all knowledge about society. On the one hand, "human being" is too much because the various sciences have, by observing from different perspectives, produced so much knowledge about human beings. We have accumulated valuable information, for instance, from the perspectives of biology, chemistry, and neurophysiology; from the perspectives of psychology, economics, and political science; and from the perspectives of demography and population ecology. Additionally, knowledge has been won from taking non-scientific perspectives on human beings, such as that of literature, psychoanalysis, and the mass media. Last but not least, each human being has collected knowledge about what it means to be a human being. It would be impossible to incorporate all this knowledge into a theory of society. Our fundamental problem can be put into a single question: what information about human beings should sociology take into account and what should it leave out? How much variety can it attempt to organize without losing its own discipline? On the other hand, the concept of "human being" is too small of a reference point to support any general theory of society. Sociological theory must also address the tremendous impact of "non-human" factors such as regional food supplies, climate, the domestication of plants and animals, literacy, metallurgy, computers, genetics, disease, deforestation, fuel sources, and a kaleidoscope of technological forces that have influenced the social lives of human beings. After narrowing our viewpoint to see nothing larger than human beings, how can sociology ignore the importance of socio-cultural developments such as writing, money, art, libraries, and the internet? For that matter, sociological theory must also consider "transcendent forces" that have touched the collective experience of humans: lightning bolts, spirits, ghosts, curses, mythical creatures, and gods. Thus, in the interests of building a scientific theory of society, social system theory has decided to accept the premise that society does not consist of human beings.

According to the distinction drawn by social system theory, society is the entirety of all communication. Thus, the distinction of society indicates a clear and definite field of inquiry. Everything that realizes communication realizes (at the same time) society. Answering the question, "How do you do?" realizes society. Any possible communication that might result from writing and reading this sentence would also realize society. Everything else in the world—everything that does not realize communication—does not belong to society. A rose is a rose is a rose. The rose is not "part" of society; however, the poem by Gertrude Stein about the rose is certainly a part of it. No feeling, no thought, no heartbeat is part of society. Nonetheless, everything could become part of society, if only a form of communication handles it as a difference that makes a difference for communication. A rose and feelings and certain thoughts and heartbeats may become a communication of love, as long as the right words at the right times, and the right silences at the right times are integrated and exposed through communication. Everything outside of communication remains outside of society; but this does not mean that everything outside of communication is unimportant for society or for a theory of society.

The theory of the social system is an account of a distinction or difference: specifically, of the difference between system and environment. The environment is and will remain important on an essential level. Without an environment, there could be no system—no context against which the system could draw a boundary, organize complexity, and thereby identify itself by means of drawing and repeating a distinction. Communication cannot happen without the human hearts, nerves, and minds that constitute parts of its environment. However, as Luhmann inquired, "How can communication reproduce itself if it must rely on a multitude of nervously vibrating brains and agitated minds?" (Luhmann 2002:171).

Before we can introduce any sweeping statements about society, we must introduce one more elementary distinction: the distinction between society and communication. If, on the one hand, only communication realizes society, and on the other hand, every single communicative act realizes society: What makes society a specifically social unit? What is the difference between "How do you do?" and society?

Within the theory, every single communication is thought to refer to other communications in its environment. "How do you do?" is asked countless times by countless speakers. The same question is posed while other communications declare love or bankruptcy, while orders to kill are issued, while somebody is complaining about the burden of his work. Society has no communication in its environment, because everything that is communication belongs to its inside and not to its outside. This does not mean society is a big container with myriads of ready-made communications that can be called into use whenever required. Rather, it means that society is present in every single communication as the final nexus for deciding what can and cannot be communicated. Society situates every single unit of communication with reference to the general limits of communication. Therefore society is not a warehouse of free choice, but a filter that allows some things to pass as possible communication while withholding everything else. Among other things that it does, the filtering capacity of society is what restrains the thoughts and feelings of human beings from ever reaching the level of communicative or societal reality.

Transmitting thoughts and feelings is neither the problem nor the function of society. The problem of society is only to differentiate communication with regard to the problem of reproducing the boundary between communication and everything else. "How do you do?" contributes in a different way to the reproduction of the possibility of communication than does an order to kill, writing and reading a book, or debating the merits of a theory of society. Society "cares" for all of these differences and the possible meanings that can be connected with these differences. Society is the recursive connectivity of all communication, distinctly uninterested in everything that each concrete unit of communication "wants to say." We must underscore that this unity is an imaginary construction: it cannot be observed in an all-encompassing manner.

After asserting that society comprises the set of all units of communication, social system theory turns its attention to two related questions: how are the elements of communication connected and how is the boundary that distinguishes society from everything else reproduced? Pursuing these questions, this chapter describes four historical forms of differentiating society, along with conditions that appear to influence the evolution of such forms of connecting communications. Specifically, we discuss the selective manner in which segmentary, stratified, centralized, and functionally differentiated forms of society structure the possibilities of communication. The evolution of these societal forms directly corresponds with the evolution of communication technology. In the modern era, we can observe all four forms concurrently, but systems theorists tend to focus on functional differentiation. As modern society differentiates itself internally, it increases in complexity and places itself under the strain of having to coordinate increasingly diverse elements of communication in increasingly selective relations.

The Four Forms of Society

Taking a historical and comparative perspective, we can distinguish four significantly different types of solutions to the problem of building social boundaries and maintaining communicative connectivity. Types of differentiation can have random, unpredictable beginnings and they are not regulated by any internal laws that might guide, in a Hegelian sense, a developing form. Nonetheless, Luhmann asserted that there "seems to be a kind of selection that chooses what is capable of becoming permanent… the only forms of differentiation able to survive are those that can mobilize processes of deviation-amplification (positive feedback) to their own advantage and keep themselves from being leveled out again" (1995a:190). *Segmentary, centralized, stratified, and functionally differentiated* forms of society demonstrate a common survival strategy, in that their continued autopoiesis refers to a *primary* distinction that helps orient and restrict communication. All four forms differentiate communication as either "equal" or "unequal," but they perform this operation by employing different concepts of equality (Luhmann 1997a:613). It is important to note that the concept of "equality" does not imply egalitarianism, justice, or any inherently "positive" manner of association. It is certainly not a normative claim about how to distribute resources or respect. "Equality" is referenced only as a construct to differentiate societal forms of communication. One might also conceive of the equality/inequality distinction in terms of similar/dissimilar, as Luhmann also suggested (1995a:188). Different boundaries occur mainly in regard to the distinction of inclusion and exclusion. This primary distinction of equality/inequality decides which aspects of personal relevancies or attributes can be connected within communicative systems, and which personal relevancies are excluded as "unadressable" within a certain societal form.

SEGMENTARY SOCIETY

Segmentary societies or tribal societies give primacy to a particular form of system differentiation: society is divided into equal parts that reciprocally produce environments for each other (Luhmann 1997a:634). In its most basic form, a segmentary society involves independent families who represent the unity of a horde. Each family is a symbolic unity of members of different ages and sex, producing a unity in difference that can be grouped with similar unities at a higher level of social complexity: the tribe. As the population increases, the tribe can reproduce itself by dividing and founding new settlements. Families are unified in villages, and villages are unified in tribes according to family relations or shared territory. In a segmentary society, communication is structured by the immediate co-presence of participants who are symbolically distinguished with reference to their belonging or not belonging to a particular tribe or family. By co-presence, we mean that participants communicate while sharing location and time and that their communication is based on an intensive sharing of space and time.

Inequality within a segmentary society is regulated by simple internal boundaries. The roles of men and women are elaborated in a division of labor: chieftains are differentiated from warriors, shamans are differentiated from the faithful, children from adults, and so on. Regularly interacting face to face while struggling to solve similar problems of everyday life, men and women, chiefs and warriors, and young and old can easily anticipate one another's knowledge, concerns, and expectations. Discussing this sort of simple differentiation within "primitive society," Robert Redfield suggested that people differed in prestige and power, even while possessing an identical culture (1953:42).

Observing the low level of cultural differentiation between members of a tribe does not mean that men cannot exclude women or children from their "sacred" or "fraternal" communication. Members of segmentary societies can selectively exclude others from communication by regulating access to the site and time of special interactions. All men, for instance, might be restricted from attending a birth. As a group, women might not be permitted to attend a circumcision. No child is always welcome in his parent's tent. Such restrictions refer to the inequality of members that is equally reproduced in each family and tribe.

In his famous ethnography of segmentary societies on the Trobriand Islands (1929), Bronislaw Malinowski related many examples of restricted communication among unequals. The presence of men is expressly forbidden, for instance, when women engage in the following social activities: communal weeding, collecting shells, fetching water, gathering firewood, and making clothing (1929:493). Malinowski also described a "systematic" form of "feminine wooing" known as *kamali*, which communicates romantic expectations and selections in the form of "erotic scratching:"

> It is a general rule in all the districts of the Trobriands that, when a boy and girl are strongly attracted to each other, and especially before their passion is satisfied, the girl is allowed to inflict considerable bodily pain on her lover by scratching, beating, thrashing, or even wounding with a sharp instrument… [During the collective ritual of *kamali kayasa*, a group of] boys in gala dress would walk round the central place singing: girls would come up to them and teasing jokes and repartee would be exchanged… Women, who were expected on such occasions to be much more forward than usual, would pass from teasing to scratching, and attack the boys with mussel shells and bamboo-knives, or with a piece of obsidian or a small sharp axe. A boy was allowed to run away, and would do so if his assailant were not attractive to him. But it was a sign of manliness and a proof of success to be properly slashed about. Also, when a boy was attracted by a girl, he would, naturally, not run away, but take her attack as an invitation. The ambition of a woman was to slash as many men as she could; the ambition of a man to carry away as many cuts as he could stand, and to reap the reward in each case. (1929:493)

During *kamali*, communication takes place in a selected time and place, the roles of the participating young women and men are clearly defined, and older adults and children are excluded. Although the young men may vary in their "attractiveness," they each participate in the communication process as "equals." Communicating observations within the form of *kamali,* any of the participating young women may meaningfully *cut and indicate* any of the singing young men. It is hard to think of a better illustration of George Spencer Brown's notion of making a distinction through severance!

The relatively high probability of successful communication between co-present and equal entities in segmentary society is reduced as internal differentiation increases. As Emile Durkheim put it, "Solidarity which comes from likenesses is at its maximum when the collective conscience completely envelops our whole conscience and coincides in all points with it. But, at that moment, our individuality is nil… If we have a lively desire to think and act for ourselves, we cannot be inclined to think and act as others do" (1964:130). Participants (individuals, families, clans, villages, etc.) that surprise others with creative observations, innovations, and custom-breaking experiments will most likely face misunderstanding, resentment, and sanctions. Innovative observations threaten the established distinctions of the segmentary society, a stabilized order that depends on equally limiting the possibilities of all participants in communication. Durkheim's famous concept of the "collective conscience," which "designates the totality of social similitudes" (1964:79–80), takes on a more precise meaning if it is understood as the semantic apparatus that establishes the structural possibilities of communication within a social system. It is the available latent medium, the totality of loose couplings, from which tight couplings can contingently take form and produce meaning during real-time instances of communication.

In a segmentary society, there is a very close relationship between internal boundaries to the inclusion and exclusion of persons. Every person included in the community possesses a valid address and the cultural training generally required for participation in communication. However, when a person is excluded from participation in the internal order, he or she is consequently excluded—excommunicated—from all society.

Robert Redfield's description of the homogenous "little community" might be extended to the segmentary society:

> Activities and states of mind are much alike for all persons in corresponding sex and age positions; and the career of one generation repeats that of the preceding. So understood, homogenous is equivalent to "slow changing." [S]elf sufficient and providing for all or most of the activities and needs of the people in it, the little community is a cradle-to-the-grave arrangement. (1956:4)

In a quite literal sense, exclusion from one's segment (family, tribe, etc.) typically meant exclusion from society itself. In the absence of any social alternative, exlusion typically meant death for individuals, whatever their age, sex, or prestige.

The equal and common culture of primitive society erodes, according to Redfield, with the appearance of cities and "civilization:"

> The strongest impression is that in societies unaffected by civilization men change their ideas and their ways of doing things, and are not infrequently aware that they are doing so, but under conditions of isolation these changes are small changes. The reach into the future to make life different from what it was is a short reach. And that reach occurs when some immediate circumstances present the people with a difficulty...Civilization is breakdown of the old ways. It is a meeting of many minds. It is the weight of new exactions upon human labor; and it is the organization and mass production of food, buildings, war, cruelty, and political adventure. (1953:133, 136)

The limited possibilities for making meaning within the "common culture" of segmentary society contribute to the success of communication, but the same restrictions can build pressure for increased internal differentiation.

Centralized Society

When society is differentiated with respect to the difference between *center and periphery*, communication is organized on a higher level of complexity. One might think of a city and its surrounding countryside, an empire and its outlying areas, or a corporation's headquarters and its distant franchises. There is much more than a simple reference to one's tribe as the navel of the world, the center of the universe—a typical characteristic of segmentary groups. This second form lifts communication from the need for interaction and the sharing of space and time. The distinction between this place and that place is reproduced again and again as communication itself references the boundary of center and periphery. Communication makes sense by referring itself to the center's horizon of possible meanings, to a core set of cultured expectations and established frames.

The spatial dimension plays the most important role in this case, with social participants occupying locations at a relative distance from a clearly established and repeatedly referenced focal point. As the possibilities of communication transcend the borders of a civilization, it becomes necessary to differentiate as well as universalize the concept of human beings. On an abstract level, strangers and savages are people, but they do not share one's own sense of order, civilization, and perhaps divine election. This style of thinking is well illustrated in an article entitled "Manifest Destiny," published in 1885 by the American philosopher John Fiske:

> The work that the English race began when it colonized North America is destined to go on until every land on the earth's surface that is not already the seat of an old civilization shall become English in its language, in its religion, in its political habits and traditions, and to a predominant extent in the blood of its people… I believe that the time will come when such a state of things will exist upon the earth, when it will be possible to speak of the United States as stretching from pole to pole… Indeed, only when such a state of things has begun to be realized can civilization, as sharply demarcated from barbarism, be said to have fairly begun. Only then can the world be said to have become truly Christian. (1885:558)

Communication begins to distinguish between a preferred location such as a preeminent city (Athens, Rome, Cairo, or Constantinople) in contrast to outlying provinces or territories, differentiating thereby between center and periphery. As Luhmann asserted with regard to the center/periphery distinction (1997a:666): "The oldest model is marked by a strict separation between land that is occupied or available for occupation and a surrounding wilderness. In one's own civilized land, one may live, build, and establish cults. Here there is memory and civilization. Outside, there is a wilderness full of surprises and frightful things."

The periphery may be observed as a colonized region that is exploited for the economic gain of a central elite, but of greater significance is its capacity to supply variety to communication and foment social change. On the one hand, the unorganized chaos just outside the periphery threatens the known order of the civilized center. The ancient Greeks, for instance, used the word *barbarian* to describe those exotic people who could not understand their language: incomprehension was a source of fear and marked the savage. On the other hand, the language, food, technology, art, religion, and military strategies of barbarians on the margins of society stimulated innovation and cultural evolution in the center. While protecting and importing resources from the periphery, the center exports cultural and structural resources and encourages communicative connectivity along its own well-worn channels. To overcome the unorganized variety of vernacular languages in the provinces, dispersed elites representing the center correspond with each other using a common language. Chinese, Arabic, Latin, French, and English have all played such a role in the history of empires. In Luhmann's understanding, an empire is a "side effect of the extension of communication." Without having clear borders, an empire is the horizon of meaning that emerges from the communication of bureaucratic elites. Such elites take the exclusive status of their empire for granted and if they ever consider territorial borders, then only as temporary limits to their empire's future expansion (Luhmann 1997a:670–671).

The periphery, still clinging to the more probable, less structured segmentary differentiation of its households and clans, tends to be less invested in maintaining the center's organized complexity (Luhmann 1997a:663). Hence, the center's authority and privilege tend to require the work of cultural missionaries, backed by an enduring military presence. When discipline declines, improbable civilization goes to the vandals. Commercial, military, evangelistic, and other interests must be strong enough to motivate the hard work of maintaining the center's contingent social order while also engaging the disorder of the periphery. The elites in the center thrive on contact with the outside, but this may not be reciprocal. This form of society depends on communication between unequal parties, the civilized and the wild, which internally appear as equals.

Compared to a segmentary society, there is more social mobility in a centralized society. Individual people have more control over their own inclusion in or exclusion from communication. Individuals may move in and out of regions, take up specialized occupations, learn different languages, send and receive texts, and experience greater anonymity. Whereas segmentary differentiation strictly binds an individual's social roles into a unity of membership, centralized society permits individuals to occupy relatively independent social statuses, enhancing inequality. Nevertheless we cannot speak of general flexibility and mobility for individual members of centralized societies.

STRATIFIED SOCIETY

Stratified societies differentiate communication in yet a more complex manner, enabling different groups to emerge within the same society, with a minority dominating the rest. Each part develops its own semantic and structural possibilities, while at the same time preserving its place within an integrated social system. This form of internal differentiation maintains identity and equality within a stratum, and difference and inequality between strata. Stratified societies order their communication primarily according to unequal ranks, that is, according to a hierarchy from "top" to "bottom," with different grades of complexity. The elites in a stratified society do not view their inferiors as relatives; society is not formed by shared descent. Endogamy protects the wealth and social identity of elites, while also demonstrating a distinctly different bloodline and heritage (see Luhmann 1997a:680).

We can observe simple stratified societies comprised of dualities such as citizens/slaves or aristocrats/commoners. We can also find more complex examples of stratified societies, such as medieval Europe's feudalism or India's caste system. Otto von Gierke described how medieval thought proceeded from the idea of a single corporate unity of different ranks:

> All Manyness has its origin in Oneness (*omnis multitudo derivatur ab uno*) and to Oneness it returns (*ad unum reducitur*). Therefore all Order consists in the subordination of Plurality to Unity (*ordinatio ad unum*), and never and nowhere can a purpose that is common to Many be effectual unless the One rules over the Many and directs the Many to the goal…Unity is the root of All, and therefore of all social existence. (1987:9)
>
> From the notion of an Organism, whose being involves a union of like with unlike, was derived the necessity of differences in rank, profession and estate, so that the individuals, who were the elements in ecclesiastical and political Bodies, were conceived, not as arithmetically equal units, but as socially grouped and differentiated from each other. (1987:27–28)

Within a stratified society, equality is realized *within the strata* of the different ranks. The priest was not the duke's equal, but he was the equal of other priests and could participate in ecclesial communication.

Within India's caste system, Heinrich Zimmer asserted that the "supreme virtue" is for individuals to become "assimilated to the timeless, immemorial, absolutely impersonal mask" of the caste into which they were born:

> The correct manner of dealing with every life problem that arises, therefore, is indicated by the laws (*dharma*) of the caste (*varna*) to which one belongs, and of the particular stage-of-life (*asrama*) that is proper to one's age. One is not free to choose; one belongs to a species—a family, guild and craft, a group, a denomination. And since this circumstance not only determines to the last detail the regulations for one's public and private conduct… one's concern as a judging and acting entity must be only to meet every problem in a manner befitting the role one plays. (1974:152–153)

Thus, members of the same caste expect one another to limit communicative possibilities and selections of meaning in the same manner, reducing complexity according to the laws of their caste. In this way they become equal partners in communication, even while differentiating themselves from other castes.

According to Max Weber, rank (*staendische Lage*) primarily determines an individual's marital opportunities, commensurability with peers, occupational privileges, and traditions. Weber emphasizes that every rank is conventional, ordered according to its own rules, and self-imposes limits on its own possibilities (1972:179–180). Members of aristocratic families had to marry other aristocrats. For example, in 1936, King Edward VIII of England married an American divorcee named Wallis Simpson. English lawyers, peers, and commoners alike were not prepared to understand the utterance, "We, the king, marry a commoner." He was compelled to abdicate the throne after occupying it for only eleven months. Political communication also could only be made within the highest ranks. Members of lower ranks who claimed the right to join in political discussions were rebels or demagogues. Questions of subsistence and everyday life had to be dealt with on all strata, but could be addressed only within the same strata. As an illustration, consider the five traditional ranks comprising the British "peerage" or aristocracy: duke, marquees, earl, viscount, and baron (in descending order). From the Latin *par,* the word *peer* implies equality. In a stratified society, one communicates only with one's peers; rank determines addressability.

Functionally Differentiated Society

The last means of differentiating society references functional differentiation and deserves a closer look because it is the most complex form of society. It is also the form of society that most profoundly shapes our present experience of communication. What is the relation of "equal" and "unequal" communication in the case of the functionally differentiated society? How does it solve the problem of drawing the boundary between communication and its environment? *Functional differentiation* means that there are many specific problems that have to be solved. It is these problems—and not segments, ranks, regional preferences, or the wishes of individuals—that dictate which communication can connect to which other communication. If one is in need of goods or services, one learns to get enough money to buy them—and not to pray for them, threaten violence, promise love, or start a research project about global nutrition problems. The other way around is also certain: money will not answer one's questions about salvation and perdition, guarantee health, secure love, or answer questions of science. Connecting to functionally specific communication means selectively avoiding connections that also could have been made. Functionally specific communication disconnects different problem types, and it disconnects from all communication that remains functionally unspecified. The question, "How are you?" does not point out which problem—if any problem at all—might be solved by the conversation that follows. The modern solution of differentiating between "equal" and "unequal" communications requires that every functionally specific communication is equal in the regard that it solves a societal problem. The communications are unequal because they solve "different" problems.

Traditional descriptions of society tend to recognize only two important functional systems: economics and politics. Consider, for example, how society was defined in a heavily marketed undergraduate sociology textbook: "A society is the population that shares the same territory and is bound together by economic and political ties. Often the members share a common culture, but not always" (Brinkerhoff et al. 2008:32). Furthermore, many theories conceptualize the relationship between economics and politics in terms of stratification. The theories suggest, for instance, that the economy is the most important system, taking the top rank of society, subordinating everything else. Politics merely sets the "frames" in which rational economic activity occurs. In this simple picture of society, science and education fulfill only "service functions" for the economy: the functions of delivering profitable technologies and capable workers. With a focus on differentiation and an orientation toward functional operations, we can gain a richer, more complex image of modern society and its internal order. This focus lets us identify a group of problems, every one of which must remain addressable in order to support the parallel, polylinear communicative connectivity that we expect from modern society. We can certainly identify an economic function: the problem of establishing present expectations about future access to goods and services.[1] But there are other functions, too, all of which must be fulfilled at the same time. It is in this sense that Armin Nassehi (2006) described modernity as the "society of multiple presences." We cannot establish a "ranking" of presences from "most important" to "least important."

Under the influence of pioneering thinkers such as Adam Smith and Karl Marx, many sociologists assume that solving economic problems automatically solves other major social problems. The fundamentalist Taliban in Afghanistan, of course, would like to use religious solutions to solve all other social problems. Modern society has too many concurrent problems to justify putting faith in a single dominant functional system. For instance, the economy cannot take over religion's efforts to answer questions about existential uncertainty, theodicy, cosmology, eschatology, salvation, and what God expects from humanity. In every single work of art, the system of art shows how the process of observation leads to order and thus to exclusion. The economic value of a painting says nothing about the impressions it evokes as a work of art. The system of law stabilizes expectations that in fact are or could be violated, but should nevertheless guide future actions (see King and Thornhill 2003). We do not expect to be killed when we run off to the store to buy milk, even though the mass media shows us that people are killed on the street every day. Politics functions to make decisions collectively binding, even for those who did not agree or will not agree to them. Science controls the admission of new knowledge into the canon of established knowledge.[2] Communication in the family must take care of the complete personalities of all its members. Members of a family share more than economic interests; they worry about maintaining good grades in school, folding laundry, supporting moral standards, making sure immunizations are current, encouraging kids not to talk back to parents, and motivating everyone to go to bed on time. With so many intimate and mundane concerns, it is clear why families must limit their size. Education, in families, as well as in formal educational institutions, has to observe and correct the self-socialization of individuals. This mainly implies the self-socialization of children, but it also involves the life-long learning of adults. The health system not only cures illnesses, but also informs everything that refers to the distinction between ill and healthy. All at once, society pursues so many different problems!

What makes the understanding of modern society and the orientation within it so confusing is the fact that there is no logical order between all these functions—no rank of relative importance, no moral ascendancy, no fundamental order in time that could point out which problems should be solved first, second, and third. All problems have to be addressed as equals, and they must be addressed simultaneously. Laws must be respected even if science has not yet determined that they are effective. Students must pass their current exams even if educational fads might eventually change the curriculum. The truth-value of a theory of evolution cannot depend on the worldviews of candidates in the current presidential elections. Different communication systems have evolved because all of these problems must be handled *now*. As Luhmann described it, each functional subsystem "reduces the complexity of its environment by assuming that many of the reproductive requirements needed in the overall system are fulfilled elsewhere by other subsystems" (1995a:192). The creation of system boundaries within society makes it possible for functional systems to selectively work on solving their own particular problem. Each functional system internally processes its meaningful connections, while on the outside other systems manage their own affairs. We participate in an uncoordinated, "polycontextural" social world (see Gotthard 1979), despite the fact that functional systems offer some fragmented islands of rather improbable coordination.

Human Beings and Functional Differentiation

The ability to make sense in a functionally differentiated society is determined by a specific structural context that is subject to historical change.

At any given time, what makes sense in politics (winning or losing an election) may not make money in the world of finance. What makes sense in school (passing or failing courses) may not produce a winning football team. Each societal system is a self-determined, autopoietic system operating in real time with its own current forms of reference. From the perspective of social system theory, as we have explained in previous chapters, there are three general kinds of communication media: Language (*Sprache*), Disseminating Media (*Verbreitungsmedien*) and Success Media (*Erfolgsmedien*). As Luhmann asserted, "Disseminating media determine and increase the number of recipients of a communication" (1997a:202). Writing, television, and email are examples of dissemination media; each can be used to pass along the same information among participants in society.

As a consequence of evolution and increased social differentiation, new types of communication media are institutionalized: success media. As we explained in chapter five, success media are symbolically generalized media that convey meaning *within a specific societal system*. A specific binary code produces the meaningful form (difference) within each success medium (Luhmann 1997a:359). For instance, money is used in the economy (income/loss), grades are used in school (pass/fail), and power is used in politics (win/lose). As a "stratified" society evolves into a functionally differentiated society, the recursive complexity and semantic differentiation of success media increases (Luhmann 1997a:205).

The term "functional differentiation" describes the *internal* order of the modern form of society, establishing possibilities for communicative connectivity. Outside the superset of all communication we find the most important environmental feature of society: *consciousness*, the mental activity of individual human participants. No single mind or psychic system could ever observe and understand the simultaneous communicative operations of the many functional subsystems. To make society consciously observable and controllable, socio-cultural evolution has invented professions and organizations. For both their members and clients, these institutions determine which problems must attract attention in which situations, and which problems must be systematically neglected. Calling one's physician, one expects to receive medical advice or a prescription, but not a spiritual message about redemption. Calling for technical assistance with one's computer, one can expect to discuss computer problems and solutions. The hardware specialist on the line might pose questions about politics, ask for a recipe for dinner, offer advice about how to become a better lover, or simply try to return to his video game: all of which would frustrate one's cultured expectations about how hardware specialists should operate. To handle our personal needs, we have been trained to seek out specifically differentiated channels of communication. Nevertheless, all problems and all functionally oriented communication potentially affects all individuals all of the time. This is because the ongoing operational continuity of society constantly adjusts the status of relevant communicative systems, thereby changing the general structure of risks and opportunities.

Individual humans win and lose positions according to the selective operations of the functional subsystems. As we observe the ongoing operations of functional systems, we notice that our own internal state is adjusted. Our favorite baseball team loses a game and our mood is spoiled; they win a championship and we bask in reflected glory. One sends a proposal and it is accepted or rejected. One is hired and fired. One marries and divorces. One is arrested and released. The functional systems, on the one hand, inspect us and make their own judgments regarding future possibilities. On the other hand, we make selections based on our evaluation of available organizations, networks, and social opportunities. After receiving two job offers, one must make a selection. When one's political party loses an election or decides to change its official position on an issue, one must come to terms with the change. When living artists no longer produce works that touch one's aesthetic sensitivities or enrich one's experience as an observer, one must learn to appreciate works completed in the past. Each functional system of society coordinates its own communicative elements. However, an individual's conscious awareness and evaluation of all functional matters or personal concerns is not socially coordinated. In this sense, the internal complexity of every individual produces the unique unity of consciousness.

The concept of "career" expresses this combination of self-selectivity and selections by communicative processes (see our discussion of "career" in chapter 6). A career marks both the participation of individuals in society *and* the sensitivity of society for an individual's ideas, requests, contributions, and complaints. Thus the form of one's career has two sides, our own selections (where to study, which skills to develop, for whom to work) and selections that concern us (admission to a college, a job offer, a promotion). This form structures the boundary of modern society, the "gate-keeping" of communication in regard to its most important environment: the consciousness of participating psychic systems. As seen from society, the career regulates the inclusion and exclusion of individuals. This regulation occurs at the operational level, one communicative element at a time, deciding who can speak, listen, sign, read, or judge. The career also regulates inclusion at the structural level, determining membership, status, or position as a whole. One could criticize this rigid selection structure and its often-disastrous effects. Is it fair that an automobile company decides to close down fourteen of its factories? Is it just that my son's application to that college was rejected? Rejections hurt and acceptance does not always appear well-grounded. One should keep in mind, however, that boundary problems must be solved somehow, and that the modern solution of the career has established a kind of complexity, efficacy, and accompanying degree of individual freedom that was unknown and inconceivable before.

The decision to exclude human beings from society allows us to identify a type of empirical operation that realizes the production and reproduction of society: communication. To identify such an operation is a precondition for applying a theory of social systems. The alternative would be to decide arbitrarily what features should belong to society and what features should not belong to society. If society consists in some sense of human beings, then what part of human beings makes the expressly social difference?

Founding our notion of society on the operation of communication, and excluding everything else, also eliminates a priori values and moral forces from the analysis of society. Communication, in this theory, does not serve to complete previously existing essentials, relations, or values. Society does not form connections between otherwise self-sufficient norms, values, groups, and individuals. Within the "old European" tradition of social philosophy, morality and ethics were thought to provide meaning throughout society. In a functionally differentiated society, however, morality is meaningless in the same way success is meaningless when applied outside of its own self-referential context. Luhmann wrote:

> In a modern society, in a society with fully developed symbolically generalized media, there is no supermedium capable of lending a uniform meaning to all communications. Once again, one may think of morality (some say: ethics). Yet the attempt to cure all of the morally weak areas of society with ethics (with a moral reflection) borders on the ridiculous. (1997a:359)

By underscoring that communication is the operation that realizes society by reproducing its own boundary, we conclude that everything else that gains a social reality and existence—including morality and ethics—must first go through the filter of communication. This holds true for work, norms, values, individuals, America, Europe, God, or whatever might be an important source of identity for an observer. This manner of studying society, this focus on the empirical operations of named observers, leads to a measured, parsimonious, and reserved style of thinking. As we mentioned at the beginning of this chapter, our decision to restrict the definition of society to communication had the purpose of enhancing the scientific status of sociology. For all that it sacrifices by abandoning man, God, and normative concerns, a "sociology of society" ends up gaining an object that can be observed. If nothing else, social system theory shows that developing a scientific theory of society is possible.

ENDNOTES

1. The possession of money symbolizes exactly this structure: I have the money *now*. And the meaning of "having now" is a moment in the future (the next moment, tomorrow, in three years), in which I buy a good or pay for a service I will need *at that future moment.*
2. The values "true" and "untrue" symbolize this selection procedure: "true" means incorporation of the new knowledge; "untrue" means tested and rejected.

Chapter 8

The Functional Method

Social System Theory and Empirical Research

To what extent is system theory capable of informing and supporting empirical research? Niklas Luhmann has been criticized for producing a grand theory without ever engaging in empirical work (for example, see Esser 2007). Skepticism about the usefulness of system theory for research purposes puts the future of the theory as a scientific endeavor in jeopardy, while also challenging the professional reputation of sociologists working to develop the theory. In these final pages, we want to evaluate the merits of this critique. To pursue this goal, we must first change our perspective. Our book has focused on explaining how society may be observed from a system theoretical vantage point. Now, however, we take the perspective of society and its functional system of "science," selecting system theory itself as the object of observation.

Empirical research is a crucial aspect of the emergence, the history, and the self-understanding of modern science. It relies on experimentation and carefully documented and controlled observations. It is empirical research that differentiates scientific knowledge from all other types of knowledge that have currency in society. From the point at which sociology successfully marked off its own subject matter (in particular, from the historically older field of psychology), the discipline has sought to establish itself as a subsystem of the societal system of science. The evolution of sociology as a science may be directly associated with the accumulation of an elaborate repertoire of methods used for collecting *the data* required for discovering and proving knowledge. Continuing up to the present, so much effort has been spent on developing methods that students beginning their study of sociology might come to the conclusion that data and methods define the core or the essence of the field. Indeed, it might appear that the scientific quality of sociology is *exclusively* defined through its methods and methodologies. This impression finds additional support from the fact that career advancement for individual sociologists is increasingly coupled with the successful attainment of projects, grants, fellowships, and other forms of competitive research assignments sponsored by a wide variety of organizations. The collection, organization, and processing of data and the demonstration of sophisticated analytical techniques (methods, computer programs, mathematical models) have, especially within the communicative context of organizationally funded projects, enormous performance advantages over projects of a theoretical nature. Nobody doubts that gathering data is hard work that requires a great deal of time; time that can typically be calculated in advance and neatly presented in a proposal. On the basis of such proposals, resources may be allotted to researchers according to rational guidelines and payments may be distributed and justified. The need for data is taken for granted, and this helps the scientist provide a plausible account for anticipated expenses. Sponsoring organizations appreciate researchers who can provide predictable timelines and calculated budgets, and favor is shown on those who can describe the data they require and the methods they will use for collecting and analyzing it. In addition, researchers with institutional identities and affiliations with governmental and non-governmental bureaucracies have a professional interest in conducting research that has an immediate application. As a consequence of these developments, contemporary sociological research is almost entirely practiced in the form of empirical research. Those who desire to develop their careers within such an empirically focused scientific milieu can expect to feel pressure to conform to its structural guidelines.

To a great extent, Luhmann was able to perform his research independently of scientific bureaucracies because his career unfolded according to the classical academic model. He taught at a university that paid him to both instruct his students and perform creative and independently directed research. He did not have to apply for grants and seek external funding for his projects. Nonetheless, the conclusion that system theory is in principle non-empirical, or distanced from empirical data, is false and can be flatly rejected after a closer, *empirical* examination. One may only accept this conclusion if one believes that the current and established state of research methods will not further evolve, that there cannot and need not be any methodological innovations in the future. Those who criticize sociological work strictly because it seems to lack an easily recognized empirical focus enjoy the support of a self-satisfied methodological establishment. The establishment, however, could not prevent Luhmann's research and that of subsequent system theorists from rocking sociology's dominant *theoretical* assumptions.

There have already been several attempts to integrate a system theoretical approach with existing methods of empirical research, most notably in Germany (for instance, Schneider 1995; Nassehi and Saake 2002; Vogd 2005; Lee and Brosziewski 2007). Commentators have demonstrated that system theorists have successfully employed familiar methods for empirical work and that the theory has stimulated innovative thought about methodological concerns. For the most part, attention has focused on qualitative methods that have themselves generated significant theoretical discussion: ethnography, interviewing, conversation analysis, discourse analysis, and the content analysis of documents. Despite their value, attempts to integrate system theory with existing methods have failed to question the standard or program for evaluating the informative value of empirical investigations. A respectable project involves the accumulation of an impressive assembly of data and demonstrates the use of one popular methodology or another. It is certain that the methodological establishment prefers observing empirical data to sifting through theoretical concerns. But what difference, from the perspective of sociology (in specific) and science (in general), does a particular "analysis" of any given set of data make? Apparently, questioning the identity of the observer of data is not worthwhile. Asking about the conditions of observing and the strategies used by the observer to recognize a datum and to selectively relate it to other data in an informative, scientific manner is also not worth the effort. When it comes to empirical research, it seems that the sociology is in the methodology and the data is in the data.

We are taking a different approach. We accept the basic rule of social system theory: the first thing a sociological researcher must do is to identify and observe an observer. One must concentrate one's attention on determining the distinctions an observer uses to make and organize observations that can yield information. What distinctions are in play when science engages in empirical research and what distinguishes such operations from other endeavors, such as theory? We must answer this question before we can decide if system theory is capable of supporting empirical research, if it can incorporate the form of observing that is described as empirical with its own observations of society and other social systems.

The central distinction used in empirical research can be quickly identified. It is none other than datum and variable (see Brosziewski 2003:136–140). A datum is, with regard to its own form, a selection from a horizon of other possible data. The traffic light is *red*, not yellow or green. The temperature outside is 67 degrees, not 23, 48, or 100. The whole horizon of possible data represents variety: it is a realm within which a datum can be recognized as a variable. A variable can be characterized by many different properties: it can be measured (temperature, for instance), it can be structured as ordinals (low, high, highest), and it can be organized as a classification scheme (the traffic light's different colors or “male or female” for the variable of gender). This form of observing is not only relevant for science. It plays a vital role in everyday life (the train leaves at 8:30 from platform 4), in journalism (politician X said Y), and in many other fields; this is increasingly so in situations where machines are used to process data and manage information (Brosziewski 2002b). It may be difficult to develop an initial sense of the difference between data and information, but drawing a distinction between datum and variable makes this plain. It should be evident that the datum offers no information by itself. A datum represents the determination of a previously open possibility (it could have been 23, 48, or 100 degrees). Information is produced when one observes the *relationship* or the *difference* between the datum and the variable. The datum is information only when, within the perspective of an observer, it is a determined variable that could have taken on another form. In this sense, the variable serves as a medium in which different data may take temporary form. To make sense out of any actual datum, an observing system must be able to relate its determined form to the horizon of other possibilities.

To put it concisely, data distinguish their variables. Therefore, in the same instant, data also distinguish between different variables. 57 degrees is not a possible color, and red is not a possible temperature. Keeping in mind the difference between datum and variable is critical when it comes to the observation of empirical research. However, we must take a further step and inquire from what the two-sided form of datum/variable is itself distinguished. Only by considering this matter can an observer access the context in which the form of datum can reveal information. Variables distinguish themselves from *constants*. This is very clear when it comes to mathematical formulas, in which variables and constants are explicitly separated. One of the most famous formulas, of course, is Albert Einstein’s $E=mc^2$; in which “c” represents a constant value (the speed of light in a vacuum) that does not change regardless of how large “m” (the mass) is (with the value of “E,” energy, directly dependent on “m” and “c”). Yet, even when the parameters are not so distinct, they still play an essential role. In principle, the names we call ourselves, for instance, are also variables. Yet one uses personal names under the assumption that the people who are called by them will remain the same; that is, that they will stay constant, even as communicative situations change.

The distinction between variable and constant is truly classic, with roots stretching back to Aristotle's metaphysics (with its concept of movement, *kinêsis*) and traditional theology (God as the unmoved mover, creation as the changing expression of an unchanging divine plan). Scientific methods are devoted to developing technical applications for relations between variable and constant values. Every use of method depends on the specification of this kind of relationship. A datum can only be determined if its range of variability is known and its value is related to other values that are acknowledged as constants. If everything changed at the same time, nothing could be determined. The instrument for taking measurements, as well as the observer who reads its changing values must both, in the end, remain stable constructions.

It might seem that these requirements only applies to quantitative methods, as if qualitative approaches to empirical research might operate under different conditions. This is not the case. It is true that quantitative data are unique in that they can be much more precisely labeled and the means of measuring and analyzing values tend to be more sophisticated from a technical point of view. Nonetheless, qualitative data are also data selected from a determined horizon of possible variants. The parameters of this horizon are typically not precisely established. Qualitative methodologies also require one to accept assumptions about constants. As an illustration of this point, we may consider the procedures involved in conversation analysis (Psathas 1995). This method examines detailed transcripts of verbal exchanges between participants engaging one another under "natural" conditions. Conversation analysis classifies "organized patterns of stable, identifiable structural features" that appear within such a transcript, "identifying principles that underpin the sequential organization of talk, the local management of turn taking, and practices relating to opening, sustaining, and closing orderly sequences" (Holstein and Gubrium 1994:266). The method can only identify and analyze its demarcated fragments of text as "data" if it can anticipate a determined range of variants that include openings, turn taking, subject changes, closings and so on. Conversation analysis, just as any quantitative analysis, depends on making *comparisons* between any given, actual, or realized possibilities and possibilities that might have also occurred. Thus, it is safe to assert that conversation analysis must also differentiate between data and variables, and that it must also refer to constants. It is assumed that *every* conversation must perform its beginning, choose and change its themes, structure and enact turn-taking, and bring about and deliver its conclusion. Additionally, connections between the structures of conversation and other social factors must be described as *stable* and *constant*: the social status of participating persons, the organizational frame of particular discussions, the number of participants, possible alliances and divisions between members, and the like. In the absence of such assumptions regarding constant relations, it would not be possible to recognize when a conversation analysis successfully brings forth knowledge.

The same conditions apply to all other qualitative methodologies. Hermeneutic methods interpret texts (broadly defined) in terms of their *possible* meanings. The text is the datum in that it selects a meaning from a repertoire of other possibilities. Furthermore, every hermeneutic approach must accept constants and their implied restrictions. In the hermeneutic tradition, constants are collectively described as "the context," which normally includes the biographical situation of the author. A text varies meanings within a constant context. When the context changes, the text changes its meanings; for instance, when one discovers that a given text does not actually come from the author one presupposed, but from an anonymous writer.

When it comes to ethnography, the actions and communicative contributions of participants in social situations are observed (see Brosziewski 2008). The assumption is that such action "defines" the situation. It is the action observed, therefore, that selects situations from a repertoire of possible situations. Once again, we detect the relationship between data and variables described above. In this case, the constant appears in the assumption that situations and the single actions that comprise them are not freely invented by participants. On the contrary, they must construct their contributions from *culture*, a resource that remains available even as situations change. Action, situation, and culture: this is the ethnographic version of datum, variable, and constant.

We may now make some significant generalizations: data differentiate themselves from variables and both of these differentiate themselves from constants. The threesome of datum/variable and constant can be described as the *form of the datum*. Empirical research represents observation in the form of the datum. The development of methods implies work on the technical uses of this form. Methodology systematizes experiences gained from such work and, most importantly, determines the relationship between data and theory. For example, methodology establishes the conditions under which forms of data may be used to discover or confirm theoretical assumptions. It is methodology that announces when conclusions (inductions, deductions, abductions) from data for theory or vice versa may be allowed, required, or forbidden. There is no possibility that the empirical itself, representing a "pure" form of data observation, could replace theory.

At this point we have the tools to test and respond to the thesis that system theory is unable to inform and support empirical research. We admitted above that system theory, as a theory, has had a difficult time applying established methods, despite some notable exceptions. Now we can provide a clear reason why this is so. System theory articulates the relations between data, variables, and constants with fundamentally different concepts than those used in established methodologies. Thus, the theory creates other forms of the form of the datum. The unities that are used in scales, rank orders, and classifications prompt us to ask about the systems that construct them. System theory cannot simply accept them as objective units that can be immediately integrated in its own applications. The case is the same for the distinctions used in qualitative methods; that is, for meaning, text, and context, and for action, situation, and culture. Before it can accept any assumption about data, variables, and constants, system theory must account for the system reference which constructs the perspective from which differences can appear and function as unities.

It is only after unities are constructed that one can ask if they are processed by the observing system as variables or constants. Any methodology that makes such decisions without clearly accounting for system reference is, for system theory, unable to claim any relevance or authority.

After all of these considerations, it remains true that system theory does have a preferred methodology with which it works with data; testing, modifying, correcting, and further developing its own theoretical constructs. System theory makes use of the *functional method* (Luhmann 1995a:52–58). This functionalism, it must be noted, has nothing to do with earlier varieties of structural or causal functionalism. The concept of function is defined as the form of a very different distinction. The concept of function indicates the *unity in difference of problem and solution*. Scientists seem to enjoy suggesting ideas about the origins of problems: the problems of individuals, the problems of politics, the problems of the economy, the problems of science, and even the problems of society. For its part, system theory suggests that this is exactly the place to work empirically. Problems must be formulated. Depending on the way in which they are formulated, problems take on different manifestations and are associated with different solutions. The functional method assumes that observers will operate under conditions that are characterized by complexity, variety, and contingency. Trivial problems have their own clearly identified solutions (2+2=4, for instance). In contrast, the non-trivial problems faced by observers entail selectivity that necessitates the recurrence of uncertainty and instability with each operation. The functional method examines the selectivity of problems and the selectivity of solutions, assuming that every given combination represents a contingency on each side.

Luhmann conducted empirical work using the functional method, creating an enormous collection of data that he was, despite his incredible productivity, unable to exhaust. At the University of Bielefeld, the institution where Luhmann worked, one may still find his legendary *Zettelkasten* (box of notes), an archive containing an as yet uncounted quantity of note cards on which Luhmann recorded both his own ideas and the results of his wide ranging research on texts and archives (Luhmann 1992;

Zorn 2003). The box of notes documents formulations of problems, solutions to problems, and orientations toward problems. This resource enables one to comparatively explore which problems were formulated at which historical periods and in reference to which social system, and what solutions were considered to address these problems. To get a memorable impression of just how Luhmann worked with empirical data, one may read his study of "Love as Passion" (1998). In this work, Luhmann provided a plethora of documents that describe problems of love that appeared over the course of socio-cultural evolution. Creatively exploiting his data base, Luhmann was able to describe the ways in which love appeared to solve a variety of different societal problems.

Thus, the empirical field of problems, solutions, and system references controls and restricts system theory in the same fashion as other forms of the empirical limit their theories. The system theorist is anything but free to issue assertions about the existence of problems. One must consult the empirical field of problem relations that can be or have been documented. Furthermore, every allegation of a problem, when confronted with the exacting standards of the functional method, becomes itself a problem. Claims of problems must prove themselves within a *double network*: the network of theory *and* the network of the empirical.

The distinction between problem and solution realizes the two-sided form that we previously identified as the hallmark of empirical research. It draws a distinction between variables and their actualizations (data) on the one side, while positioning constants on the other. If one intends to investigate the relationship between a given problem and solution, one must decide to reference one of these figures as *constant* and the other as *variable*. One normally indicates the problem as a given (constant), and then searches for a variety of possible solutions for the same problem. In this sense, documentation of actualized solutions represents the data of system theory. One may, however, also pursue the opposite tact, the way in which Luhmann preferred to work. One indicates a given phenomenon (such as romantic love) and then searches for various problems that the phenomenon appears to solve. In other words, the constant solution is treated as a unity (love) and juxtaposed against a plurality of possible problems. Working with system theory, it does not matter how the constant and the variable are positioned with respect to the schema of problem and solution. However, the theory must assume decisions about such positions before the functional method can be placed in motion, motivating and structuring the study of data (documented solutions of problems). If the methodological establishment fails to acknowledge this research process as empirical, it reflects the narrowness of its conceptual apparatus and its tendency to acknowledge only those methodological possibilities with which it is already familiar.

The functional method dovetails tightly together with system theory and, in particular, the theory's analysis of modern society. It is precisely from these close connections that the theory takes its own constants, its theoretical as well as its methodological fundamentals. In Luhmann's words, "Problems exist only as problem-systems (or as systems-problems)" (1995a:53). It takes some kind of given entity or unity to construct any *relationship* between a problem and solution or vice versa. This unity cannot be allowed to escape analysis or, as is so often the case in traditional research, be reduced to the analytical powers or intuition of the individual scientist. It takes a unity to recognize a problem as a problem and determine when a problem is solved. By acknowledging this need to observe such a unity's efforts to relate its own problems and solutions, researchers gain the ability to distinguish between "real" problems and their own imaginary constellations.

System theory pursues a single means to describing such a unity: there must be a system that accomplishes two feats with its operations. First, it draws a border separating itself from its environment, producing its own unity. Second, through the same operations, it constructs *other* unities (identities) that can reveal themselves to the system *in different forms*; here in the form of a "problem" and there in the form of a problem that has been "resolved." The functional method and its schema of problem and solution point to the empirical and theoretical aspects of system theory. This initially supports the conclusion that the method is theoretically adequate.

Additional support is found in the functional method's analysis of modern society. The thesis is, as it was empirically examined by Luhmann in numerous investigations, that modern society primarily differentiates its communication according to a variety of distinct problem orientations (see our discussion of "functionally differentiated society" in chapter 7). For the unity of modern society, the economy solves the basic problem of coping with scarcity. For modern society, politics solves the basic problem of making collective decisions binding. For modern society, science solves the basic problem of proving new knowledge claims. Within these and other functional subsystems, there can never be only one solution to basic problems. Instead, a constant variety of different kinds of solutions are designed, practiced, tested, and differentiated. We may go so far as to claim that modern society itself makes use of the functional method. It is only in this way that it can allow and enable problem orientations of every kind, even those of scientists. Society can allow this because it simultaneously ensures that most of its problems are differentiated and distributed among its functional systems. From this process, no exception is made for science. Extending Luhmann's assertion quoted above, we may conclude that problems only exist within society. Society is the quintessential problem system. Therefore, the functional method is not only theoretically adequate, but it is also adequate for society.

In light of these considerations, the functional method is by no means immune to criticism. It cannot present itself as the only possible method. Such a claim already violates a crucial assumption of system theory: namely, that all structures of modern society are in place as contingencies and that they are effective only because they can be changed. The structures of modern science and all scientific methods cannot be exempted from this encompassing law of modern society. However, the functional method certainly sets a high standard for those who would offer criticism or innovations; a higher standard, in fact, than the established methodologies are able to formulate. The standard is that any alternative method and every other empirical approach must at least perform as well as the functional method. Furthermore, an alternative must in certain aspects also perform better; they must satisfy any inadequacies attributed to the functional method. Only under such conditions would a substitution make sense.

The methodology of system theory certainly does not make working on conventional research projects any easier. If the plausibility and usefulness of research proposals are strictly evaluated in terms of the traditional guidelines established by bureaucratic funding agencies, then system theoretical projects can be expected to be deemed unworthy. Yet system theory has the potential to describe and understand phenomena in a more complex, exciting, and interesting manner. At the same time, the theory also offers very useful and illuminating simplifications of strategically significant identities and relations. Struggling in the thicket of research questions, methods, and funding problems, it can be very helpful to know that every empirical investigation is concerned with one single problem: to organize the assembly of data, variables, and constants in a valuable form. Combining theoretical, methodological, and some rhetorical abilities, it should be possible to satisfy not only oneself as a researcher and the particular interests of sociology but also reviewers and funding agencies.

Niklas Luhmann provided a simple reply to enquiries from the scientific bureaucracy. One may read it in the forward to his theory of society: "When I joined the faculty in sociology at the University of Bielefeld when it was established in 1969, I found myself confronted with the obligation to name the research projects I was conducting. Then and now, my project is described as follows: Theory of Society; Time frame: 30 years; Costs: none" (1997a:11). He calculated his efforts with precision: *Die Gesellschaft der Gesellschaft* (The Society of Society) was published in 1997. Whoever wants to take seriously an empirical approach to the articulation of problems can find insight in this monumental book: observing society is a theoretical *and* empirical endeavor.

References

Abercrombie, Nicholas, Stephen Hill, and Bryan S. Turner. 2006. *The Penguin Dictionary of Sociology*. New York: Penguin.

Alexander, Jeffrey C. 1985. "The 'Individualist Dilemma' in Phenomenology and Interactionism." Pp. 25–51 in *Macro-Sociological Theory*, edited by S. N. Eisenstadt and H. J. Helle. London: Sage.

———. 1998. *Neofunctionalism and After*. Hoboken, NJ: Wiley-Blackwell.

———. 2001. "Towards a Sociology of Evil." Pp. 153–172 in *Rethinking Evil: Contemporary Perspectives*, edited by Maria Pia Lara. Berkeley: University of California Press.

Alexander, Jeffrey C., Bernhard Giesen, Richard Muench, and Neil J. Smelser. 1987. *The Micro-Macro Link*. Berkeley, CA: University of California.

Aziz-Zadeh, Lisa, Stephen M. Wilson, Giacommo Rizzolatti, and Marco Iacoboni. 2006. "Congruent Embodied Representations for Visually Presented Actions and Linguistic Phrases Describing Actions." *Current Biology* 16:1818–1823.

Baecker, Dirk. 1999. *Problems of Form*. Palo Alto, CA: Stanford University Press.

———. 2001. *Wozu Kultur?* Berlin: Kulturverlag Kadmos.

———. 2002. *Wozu Systeme?* Berlin: Kulturverlag Kadmos.

Bateson, Gregory. 1972. *Steps to an Ecology of Mind*. New York: Ballantine Books.

Berger, Peter and Thomas Luckmann. 1966. *The Social Construction of Reality*. Garden City, NY: Anchor.

Bradfield, Bruce Christopher and Zelda G. Knight. 2008. "Intersubjectivity and the Schizophrenic Experience: A Hermeneutic Phenomenological Exploration." *South African Journal of Psychology* 38:33–53.

Brents, Barbara G. and Katheryn Hausbeck. 2001. "State-Sanctioned Sex: Negotiating Formal and Informal Regulatory Practices in Nevada Brothels." *Sociological Perspectives* 44: 307–333.

Brinkerhoff, David B., Lynn K. White, Suzanne T. Ortega, and Rose Weitz. 2008. *Essentials of Sociology*. Belmont, CA: Wadsworth.

Brosziewski, Achim. 2002a. "Beobachtungen der Moderne in der Systemtheorie von Niklas Luhmann." Pp. 99–125 in *Theorien der Gesellschaft*, edited by Christian Lahusen and Carsten Stark. München: Oldenbourg.

———. 2002b. *Computer, Kommunikation und Kontrolle. Eine Fallstudie zum informatisierten Management*. Konstanz: UVK.

———. 2003. *Aufschalten. Kommunikation im Medium der Digitalität*. Konstanz: UVK.

———. 2008. "Die Pädagogik (in) der ethnographischen Form." Pp. 65–75 in *Ethnographie und Erziehungswissenschaft. Methodologische Reflexionen und empirische Annäherungen*, edited by Bettina Hünersdorf, Christoph Maeder, and Burkhard Müller. Weinheim: Juventa.

Brunsson, Nils and Johan P. Olsen. 1993. *The Reforming Organization*. New York: Routledge.

Cassirer, Ernst. 1965. *The Philosophy of Symbolic Forms: Mythical Thought*. New Haven, CT: Yale University Press.

Collins, Randall. 1981. "On the Microfoundations of Macrosociology." *American Journal of Sociology* 86:984–1014.

———. 2005. *Interaction Ritual Chains*. Princeton, NJ: Princeton University Press.

Cooley, Charles Horton. 1907. "Social Consciousness." *Proceedings of the American Sociological Society* 1:97–109.

———. 1909. *Social Organization: A Study of the Larger Mind.* New York: Charles Scribner's Sons.

———. 1922. *Human Nature and the Social Order*. New York: Charles Scribner's Sons.

———. 1927. *Life and the Student: Roadside Notes on Human Nature, Society, and Letters*. New York: A.A. Knopf.

Crelin, Edmund S. 1987. *The Human Vocal Tract: Anatomy, Function, Development, and Evolution*. New York: Vantage.

Davis, Kingsley. 1949. *Human Society.* New York: Macmillan.

Dewey, John. 1925. *Experience and Nature*. La Salle, IL: Open Court.

———. 1927. *The Public and Its Problems.* Denver, CO: Alan Swallow.

Diamond, Jared. 1992. *The Third Chimpanzee: The Evolution and Future of the Human Animal.* New York: Harper Perennial.

———. 1999. *Guns, Germs, and Steel: The Fates of Human Societies.* New York: W. W. Norton.

Dreeben, Robert. 1968. On *What is Learned in School*. Reading, MA: Addison-Wesley.

Durkheim, Emile. 1938. *The Rules of Sociological Method.* Glencoe, IL: The Free Press.

———. 1964. The Division of Labor in Society. New York: Free Press.

Eisenstein, Elizabeth. 1979. *The Printing Press as an Agent of Change: Communications and Cultural Transformations in Early-Modern Europe*. Cambridge: Cambridge University Press.

Enticott, Peter G., Patrick J. Johnston, Sally E. Herring, Kate E. Hoy, and Paul B. Fitzgerald. 2008. "Mirror Neuron Activation Is Associated with Facial Emotion Processing." *Neuropsychologia* 46(11):2851–2854.

Esser, Hartmut. 2007. "Soll das denn alles (gewesen) sein? Anmerkungen zur Umsetzung von soziologischer Systemtheorie in empirische Forschung." *Soziale Welt. Zeitschrift für die sozialwissenschaftliche Forschung und Praxis* 58:351–358.

Fernald, Anne. 1992. "Human Maternal Vocalization to Infants as Biologically Relevant Signals: An Evolutionary Perspective." Pp. 391–428 in *The Adapted Mind: Evolutionary Psychology and the Generation of Culture*, edited by Jerome H. Barkow, Leda Cosmides, and John Tooby. New York: Oxford University Press.

Fiske, John. 1885. "Manifest Destiny." *Harper's New Monthly Magazine* 70:578–590.

Fleischmann, Eugène. 1964. "De Weber à Nietzsche." *Archives Européennes de Sociologie* 5:190–238.

Fogassi, Leonardo and Pier Francesco Ferrari. 2007 "Mirror Neurons and the Evolution of Embodied Language." *Current Directions in Psychological Science* 16:136–141.

Freud, Sigmund. 1987. *Die Traumdeutung*. Frankfurt am Main: Fischer.

Fuchs, Peter. 1999. *Liebe, Sex und solche Sachen*. Konstanz: UVK.

———. 2002a. "The Time of Communication." Unpublished manuscript.

———. 2002b. "Die konditionierte Koproduktion von Kommunikation und Bewußtsein." Pp. 150–175 in *Ver-Schiede der Kultur, Aufsätze zur Kippe kulturanthropologischen Nachdenkens*, edited by the Arbeitsgruppe "menschen formen" of the Institute for Sociology at the Free University of Berlin. Marburg: Tectum.

———. 2004. *Sinn der Beobachtung*. Weilerswist: Velbrück.

Fuchs, Stephan. 2001. *Against Essentialism*. Cambridge, MA: Harvard University Press.

Gallese, Vittorio, Luciano Fadiga, Leonardo Fogassi, and Giacomo Rizzolatti. 1996. "Action Recognition in the Premotor Cortex." *Brain* 119:593–609.

Gallese, Vittorio and Alvin Goldman. 1998. "Mirror Neurons and the Simulation Theory of Mind-Reading." *Trends in Cognitive Sciences* 2:493–501.

Gallese, Vittorio, Christian Keysers, and Giacomo Rizzolatti. 2004. "A Unifying View of the Basis of Social Cognition." *Trends in Cognitive Sciences* 8:396–403.

Gay, Peter. 1973. "How the Modern World Began." *Horizon* 15:10–15.

Gehlen, Arnold. 1978. *Urmensch und Spaetkultur*. Frankfurt am Main: Rehberg.

Giddens, Anthony, Mitchell Duneier, and Richard P. Appelbaum. 2007. *Introduction to Sociology*. New York: W.W. Norton.

Giddings, Franklin Henry. 1896. *The Principles of Sociology*. New York: Macmillan.

Goffman, Erving. 1959. *The Presentation of Self in Everyday Life*. Garden City, NY: Doubleday.

Goody, Jack and Ian Watt. 1963. *The Consequences of Literacy. Literacy in Traditional Societies*. Cambridge: Cambridge University Press.

Gozdenweiser, Alexander A. 1913. "The Principle of Limited Possibilities in the Development of Culture." *Journal of American Folklore* 26:259–290.

Guenther, Gotthard. 1979. *Beitraege zur Grundlegung einer operationsfaehiger Dialektik.* Hamburg: Meiner, S.

Habermas, Juergen. 1984. *The Theory of Communicative Action.* Boston: Beacon Press.

Haferkamp, Hans and Michael Schmid. 1995. *Sinn, Kommunikation und soziale Differenzierung. Beiträge zu Luhmanns Theorie sozialer Systeme.* Frankfurt am Main: Suhrkamp.

Hauser, Marc D. 1996. *The Evolution of Communication.* Cambridge, MA: The MIT Press.

Hegel, G. W. F. 1977. *Phenomenology of Spirit*. Oxford: Clarendon Press.

Heidegger, Martin. 1996. *Being and Time*. Albany, NY: State University of New York Press.

Heider, Fritz. 1926. "Ding und Medium." *Symposion* 1:109–157.

———. 1958. *The Psychology of Interpersonal Relations*. New York: John Wiley & Sons.

———. 1959. *On Perception and Event Structure, and the Psychological Environment.* New York: International Universities Press.

Hogue, David. 2006. "Sensing the Other in Worship: Mirror Neurons and the Empathizing Brain." *Liturgy* 21:31–39.

Holstein, James and Jaber F. Gubrium. 1994. "Phenomenology, Ethnomethodology and Interpretive Practice." Pp. 262–272 in *Handbook of Qualitative Research*, edited by Norman K. Denzin and Yvonna S. Lincoln. Thousand Oaks, CA: Sage.

Hume, David. 1978. *A Treatise of Human Nature*. Oxford: Clarendon Press.

Husserl, Edmund. 1982. *Cartesian Meditations: An Introduction to Phenomenology*. The Hague: M. Nijhoff.

Iacoboni, Marco, Istvan Molnar-Szakacs, Vittorio Gallese, Giovanni Buccino, John C. Mazziotta, and Giacomo Rizzolatti. 2005. "Grasping the Intentions of Others with One's Own Mirror Neuron System." *PLoS Biology* 3:529–535.

Iacoboni, Marco, Roger P. Woods, Marcel Brass, Harold Bekkering, John C. Mazziotto, and Giacommo Rizzolatti. 1999. "Cortical Mechanisms of Human Imitation." *Science* 286:2526–2528.

Jung, Carl. G. 1955. "Dream Analysis in Its Practical Application." Pp. 159–182 in *An Outline of Psychoanalysis*, edited by Clara Thompson, Milton Mazer, and Earl Witenberg. New York: Modern Library.

———. 1959. *The Archetypes and the Collective Unconscious*. London: Routledge & Kegan Paul.

Kelley, George. 1963. *A Theory of Personality: The Psychology of Personal Constructs*. New York: W.W. Norton.

Keysers, Christian, Bruno Wicker, Valeria Gazzola, Jean-Luc Anton, Leonardo Fogassi, and Vittorio Gallese. 2004. "A Touching Sight:

SII/PV Activation During the Observation and Experience of Touch." *Neuron* 42:335–346.

Kieserling, André. 1999. *Kommunikation unter Anwesenden: Studien über Interaktionssysteme*. Frankfurt am Main: Suhrkamp.

King, Michael and Chris Thornhill. 2003. *Niklas Luhmann's Theory of Politics and Law*. New York: Palgrave Macmillan.

Kingma, Sytze F. 2008. "Dutch Casino Space or the Spatial Organization of Entertainment." *Culture & Organization* 14:31–48.

Kohler, Evelyne, Christian Keysers, Alessandra Umilta, Leonardo Fogassi, and Vittorio Gallese. 2002. "Hearing Sounds, Understanding Actions: Action Representation in Mirror Neurons." *Science* 297:846–848.

Latour, Bruno. 2005. *Reassembling the Social: An Introduction to Actor-Network Theory*. Oxford: Clarendon.

Latour, Bruno and Steve Woolgar. 1979. *Laboratory Life: The Social Construction of Scientific Facts*. Thousand Oaks, CA: Sage.

Lee, Daniel B. 2005a. "Ritual and the Social Meaning and Meaninglessness of Religion." *Soziale Welt* 56:7–18.

———. 2005b. "Making Music Out of Noise: Barbershop Quartet Singing and Society." *Soziale Systeme* 11:271–292.

———. 2007. "Observing Communication: Niklas Luhmann and the Problem of Ethnography." *Soziale Systeme* 13:454–465.

———. 2009. "Maria of the Oak: Society and the Problem of Divine Intervention." *Sociology of Religion* 70:213–231.

Lee, Daniel B. and Achim Brosziewski. 2007. "Participant Observation and Systems Theory: Theorizing the Ground." *Soziale Welt* 58:255–269.

Linton, Ralph. 1936. *The Study of Man*. New York: Appleton-Century.

Locke, John L. and Catherine E. Snow. 1997. "Social Influences on Vocal Learning in Human and Non-Human Primates. Pp. 274–292 in

Social influences on Vocal Development, edited by Charles Snowdon and Martine Hausberger. Cambridge: Cambridge University Press.

Luhmann, Niklas. 1979. *Trust and Power*. Chichester: Wiley.

———. 1981. "The Improbability of Communication." *International Social Science Journal* 23:122–132.

———. 1989. *Die Wirtschaft der Gesellschaft*. Frankfurt am Main: Suhrkamp.

———. 1990a. "Gesellschaftlische Komplexitaet und oeffentliche Meinung." Pp. 163–175 in *Soziologische Aufklaerung 5*. Wiesbaden: VS Verlag.

———. 1990b. *Die Wissenschaft der Gesellschaft*. Frankfurt am Main: Suhrkamp.

———.1992. "Kommunikation mit Zettelkästen: Ein Erfahrungsbericht." Pp. 53–61 in *Universität als Milieu. Kleine Schriften*, edited by André Kieserling. Bielefeld: Haux.

———. 1993. "Die Parodoxie der Form." Pp. 197–212 in *Kalkuel der Form*, edited by Dirk Baecker. Frankfurt am Main: Suhrkamp.

———. 1994. "Speaking and Silence." *New German Critique* 61:25–37.

———. 1995a. *Social Systems*. Palo Alto, CA: Stanford University Press.

———. 1995b. "Wie ist Bewusstsein an Kommunikation beteiligt?" Pp. 37–54 in *Soziologische Aufklaerung 6*. Opladen: Westdeutscher Verlag.

———. 1996a. *Die Realität der Massenmedien*. Opladen: Westdeutscher.

———. 1996b. *Protest. Systemtheorie und soziale Bewegungen*. Frankfurt am Main: Suhrkamp.

———. 1996c. "Membership and Motives in Social Systems." *Systems Research* 13:341–348.

———. 1997a. *Die Gesellschaft der Gesellschaft*. Frankfurt am Main: Suhrkamp.

———. 1997b. “The Control of Transparency.” *Systems Research and Behavioral Science* 14:359–371.

———. 1998. *Love as Passion*. Palo Alto, CA: Stanford University Press.

———. 2000a. *Art as a Social System.* Palo Alto, CA: Stanford University Press.

———. 2000b. *Organisation und Entscheidung*. Wiesbaden: Westdeutscher.

———. 2000c. *Die Religion der Gesellschaft*. Frankfurt am Main: Suhrkamp.

———. 2000d. *Die Politik der Gesellschaft*. Frankfurt am Main: Suhrkamp.

———. 2001a. “Die Paradoxie der Form. ” Pp. 243–261 in *Aufsätze und Reden*. Stuttgart: Reklam.

———. 2001b. “Dekonstruktion als Beobachtung zweiter Ordnung.” Pp. 261–296 in *Aufsätze und Reden*. Stuttgart: Reklam.

———. 2002. *Theories of Distinction.* Palo Alto, CA: Stanford University Press.

———. 2004. *Schriften zur Paedagogik*. Suhrkamp: Frankfurt am Main.

———. 2005. “Soziologie als Theorie Sozialer Systeme.” Pp. 143–172 in *Soziologische Aufklärung 1. Aufsätze zur Theorie der Gesellschaft*. Wiesbaden: VS Verlag.

———. 2006 “System as Difference.” *Organization* 13:37–57.

Luhmann, Niklas and Peter Fuchs. 1989. *Reden and Schweigen.* Suhrkamp: Frankfurt am Main.

Luijpen, William A. 1960. *Existential Phenomenology*. Pittsburgh, PA: Duquesne University Press.

Malinowski, Bronislaw. 1929. *The Sexual Life of Savages in North-Western Melanesia: An Ethnographic Account of Courtship, Marriage and Family Life Among the Natives of the Trobriand Islands, British New Guinea*. New York: Readers League of America and Eugenics Publishing Company.

Marx, Karl. 1978. *The Marx-Engels Reader*. New York: W.W. Norton.

Maturana, Humberto. 1981. "Autopoiesis." Pp. 21–33 in *Autopoiesis: A Theory of Living Organization*, edited by Milan Zeleny. New York: North Holland.

Maturana, Humberto. 2002. "Autopoiesis, Structural Coupling, and Cognition." *Cybernetics and Human Knowing* 9:5–34.

Mead, George H. 1934. *Mind, Self and Society*. Chicago: University of Chicago Press.

Merton, Robert K. 1957. "The Role-Set: Problems in Sociological Theory." *British Journal of Sociology*. 8:106–120.

Mills, C. Wright. 2000. *The Sociological Imagination*. New York: Oxford.

Mouras, Harold, Serge Stoléru, Virginie Moulier, Mélanie Pélégrini-Issac, Roland Rouxel, Bernard Grandjean, Dominique Glutron, and Jacques Bittoun. 2008. "Activation of Mirror-Neuron System by Erotic Video Clips Predicts Degree of Induced Erection: An fMRI Study." *Neuroimage* 42:1142–1150.

Nassehi, Armin. 2005. "Organizations as Decision Machines. Niklas Luhmann's Theory of Organized Social Systems." *The Sociological Review* 53:178–191.

———. 2006. *Der soziologische Diskurs der Moderne*. Frankfurt am Main: Suhrkamp.

Nassehi, Armin and Irmhild Saake. 2002. "Kontingenz: Methodisch verhindert oder beobachtet? Ein Beitrag zur Methodologie der qualitativen Sozialforschung." *Zeitschrift für Soziologie* 31:66–86.

Ong, Walter. 1982. *Orality and Literacy: The Technologizing of the Word*. London and New York: Methuen.

Owings, Donald H. and Eugene S. Morton. 1998. *Animal Vocal Communication: A New Approach*. Cambridge: Cambridge University Press.

Parsons, Talcott. 1951. *The Social System*. Glencoe, IL: The Free Press.

———. 1963. *Structure and Process in Modern Society.* Glencoe, IL: The Free Press.

Parsons, Talcott and Edward A. Shils. 1951. *Toward a General Theory of Action.* New York: Harper.

Piaget, Jean. 1960. "Wahrnehmungskonstanzen und Kausalitätswahrnehmung." *Psychologische Beiträge* 5:183–231.

Pierre, Jacob. 2008. "What Do Mirror Neurons Contribute to Human Social Cognition?" *Mind & Language* 23:190–223.

Pinker, Steven. 1995. *The Language Instinct.* New York: Harper Collins.

Plessner, Helmut. 1970. *Laughing and Crying: A Study of the Limits of Human Behavior.* Evanston, IL: Northwestern University Press.

———. 2003. *Conditio Humana.* Frankfurt am Main: Suhrkamp.

Psathas, George. 1995. *Conversation Analysis: The Study of Talk-in-Interaction.* Thousand Oaks, CA: Sage.

Ramachandran, Vilayanur S. 2000. "Mirror Neurons and Imitation Learning as the Driving Force Behind 'the Great Leap Forward" in Human Evolution." *Edge* (69). Retrieved September 4, 2009. http://www.edge.org/3rd_culture/ramachandran/ramachandran_p1.html.

———. 2007. "The Neurology of Self-Awareness." *Edge* (201). Retrieved September 4, 2009. http://www.edge.org/documents/archive/edge201.html.

Redfield, Robert. 1947. "The Folk Society." *American Journal of Sociology* 52:293–308.

———. 1953. *The Primitive World and Its Transformations.* Ithaca, NY: Cornell University Press.

———. 1956. *The Little Community.* Chicago, IL: University of Chicago Press.

Rizzolatti, Giacomo and Laila Craighero. 2004. "The Mirror-Neuron System." *Annual Review of Neuroscience* 27:169–192.

Rizzolatti, Giacomo, Leonardo Fogassi, and Vittorio Gallese. 2006. "Mirrors in the Mind." *Scientific American* 295:54–61.

Rizzolatti, Giacomo, Giuseppi Pellegrino, Luciano Fadiga, and Vittorio Gallese. 1992. "Understanding Motor Events: A Neurophysiological Study." *Experimental Brain Research* 91:176–180.

Robinson, W. Peter. 2003. *Language in Social Worlds.* Oxford: Blackwell.

Rogers, Carl R. 1951. *Client-Centered Therapy*. Boston, MA: Houghton Mifflin.

Russell, Bertrand. 1961. "Introduction." *Tractatus Logico-Philosophicus*, by Ludwig Wittgenstein. New York: Routledge and Kegen Paul.

Saville-Troike, Muriel. 1989. *The Ethnography of Communication*. New York: Basil Blackwell.

Scheff, Thomas J. 2006. *Goffman Unbound: A New Paradigm for Social Science*. Boulder, CO: Paradigm.

Scheler, Max. 1954. *The Nature of Sympathy.* London: Routledge & Kegan Paul.

Schilhab, Theresa. 2007. "Interactional Expertise Through the Looking Glass: A Peek at Mirror Neurons." *Studies in History and Philosophy of Science* 38:741–747.

Schneider, Eric C. 2001. *Vampires, Dragons, and Egyptian Kings: Youth Gangs in Postwar New York*. Princeton, NJ: Princeton University Press.

Schneider, Wolfgang L. 1995. "Objektive Hermeneutik als Forschungsmethode der Systemtheorie." *Soziale Systeme* 1:129–152.

Schulte-Rüther, Martin, Hans J. Markowitsch, Gereon R. Fink, and Martina Piefke. 2007. "Mirror Neuron and Theory of Mind Mechanisms Involved in Face-to-Face Interactions: A Functional Magnetic

Resonance Imaging Approach to Empathy." *Journal of Neuroscience* 19:1354–1372.

Schutz, Alfred. 1967. *The Phenomenology of the Social World.* Evanston, IL: Northwestern University Press.

Searle, John R. 1993. "The Problem of Consciousness." *Social Research* 60:3–16.

Seyfarth, Robert M. and Dorothy L. Cheney 1997. "Behavioral Mechanisms Underlying Vocal Communication in Nonhuman Primates." *Animal Learning and Behavior* 25:249–267.

Shadowcoven. 2006. "The Coven, My Story." Retrieved September 4, 2009. http://www.shadowcoven.zoomshare.com/1.shtml.

Shannon, Claude E. and Warren Weaver. 1949. *The Mathematical Theory of Communication.* Urbana: University of Illinois Press.

Simmel, George. 1971. *Individuality and Social Forms.* Chicago: University of Chicago Press.

Smith, Dorothy E. 2001. "Texts and the Ontology of Organizations and Institutions." *Studies in Cultures, Organizations and Societies* 7: 159–198.

Spencer Brown, George. 1979. *Laws of Form*. New York: Dutton.

Stichweh, Rudolf. 2000. "Systems Theory as an Alternative to Action Theory? The Rise of 'Communication' as a Theoretical Option." *Acta Sociologica* 43:5–13.

Stolorow, Robert. 1994. "The Intersubjective Context of Intrapsychic Experience." Pp. 3–14 in *The Intersubjective Perspective*, edited by Robert Stolorow, George Atwood, and Bernard Brandchaft. Northvale, NJ: Aronson.

———. 2000. "From Isolated Minds to Experiential World: An Intersubjective Space Odyssey." *American Journal of Psychotherapy* 54: 149–151.

Stolorow, Robert and George Atwood. 1992. *Contexts of Being: The Intersubjective Foundations of Psychological Life*. London: Analytic Press.

Sumner, William Graham. 1906. *Folkways: A Study of the Sociological Importance of Usages, Manners, Customs, Mores, and Morals.* Boston, MA: Ginn and Co.

Tenbruck, Friedrich. 1981. "Emile Durkheim oder die Geburt der Gesellschaft aus dem Geist der Soziologie." *Zeitschrift fuer Soziologie* 10:333–350.

Teubner, Gunter. 1992. "Social Order from Legislative Noise? Autopoietic Closure as a Problem for Legal Regulation" Pp. 609–649 in *State, Law and Economy as Autopoietic Systems*, edited by Gunther Teubner and Alberto Febbrajo. Milan: Dott. A. Giuffre.

Thorpe, W. H. and M. E. W. North. 1966. "Vocal Imitation in the Tropical Boubou Shrike Laniarius Aethiopicus Major as a Means of Establishing and Maintaining Social Bonds. *Ibis* 108:432–435.

Tönnies, Ferdinand. 2001. *Community and Civil Society*. Oxford: Cambridge University Press.

Tosini, Domenico. 2006. "Medium as a Basic Concept of Sociology: Contributions from Systems Theory." *Social Science Information* 45:539–560.

Turner, Ralph H. 1978. "The Role and the Person." *American Journal of Sociology* 84:1–23.

Tyack, Peter L. and Laela S. Sayigh. 1997. "Vocal Learning in Cetaceans." Pp. 208–233 in *Social Influences on Vocal Development*, edited by Charles Snowdon and Martine Hausberger. Cambridge: Cambridge University Press.

Verhagen, Arie. 2005. *Constructions of Intersubjectivity: Discourse, Syntax, and Cognition*. New York: Oxford University Press.

Vogd, Werner. 2005. *Systemtheorie und rekonstruktive Sozialforschung. Eine empirische Versöhnung unterschiedlicher theoretischer Perspektiven*. Opladen: Budrich.

von Bertalanffy, Ludwig. 1950. "An Outline of General System Theory." *British Journal for the Philosophy of Science* 1:134–165.

———. 1968. *General Systems Theory: Foundations, Development, Applications*. New York: Braziller.

von Foerster, Heinz. 1981. *Observing Systems: Selected Papers of Heinz von Foerster.* Seaside, CA: Intersystems Publications.

———. 1985. "Entdecken oder Erfinden: Wie laesst sich Verstehen verstehen?" Pp. 41–88 in *Einfuehrung in den Konstruktivismus*, edited by Heinz Gumin and Heinrich Meier. Muenchen: Piper.

———. 1987. "Erkenntnistheorien und Selbstorganization." Pp. 133–158 in *Der Diskurs des Radikalen Konstruktivismus*, edited by Siegfried J. Schmidt. Suhrkamp: Frankfurt am Main.

———. 2003. *Understanding Understanding: Essays on Cybernetics and Cognition*. New York: Springer.

von Gierke, Otto Friedrich. 1987. *Political Theories of the Middle Ages*. New York: Cambridge University Press.

von Glasersfeld, Ernst. 1981. "The Concepts of Adaptation and Viability in a Radical Constructivist Theory of Knowledge." Pp. 87–95 in *Piagetian Theory and Research,* edited by Irving E.Sigel, David M. Brodzinsky, and Roberta M. Golinkoff. Hillsdale, NJ: Erlbaum.

———. 1992. "Konstruktion der Wirklichkeit und des Begriffs der Objektivitaet." Pp. 9–40 in *Einfuehrung in den Konstruktivismus*, edited by Heinz Gumin and Heinrich Meier. Muenchen: Piper.

———. 2008. "Who Conceives of Society?" *Constructivist Foundations* 3:59–64.

Weber, Max. 1972. *Wirtschaft und Gesellschaft.* Tuebingen: Mohr.

———. 2004. *The Vocation Lectures*. Indianapolis: Hackett.

Weick, Karl E. 1969. *The Social Psychology of Organizing*. Reading, MA: Addison-Wesley.

White, Harrison, Jan Fuhse, Matthias Thiemann, and Larissa Buchholz. 2007. "Networks and Meaning: Styles and Switchings." *Soziale Systeme* 13:543–555.

Whitehead, Alfred North. 1959. *Symbolism.* New York: Capricorn.

Wicker, Bruno, Christian Keysers, Jane Plailly, Jean-Pierre Royet, Vittorio Gallese, and Giacommo Rizzolatti. 2003. "Both of Us Disgusted in My Insula: The Common Neural Basis of Seeing and Feeling Disgust." *Neuron* 40:655–664.

Zimmer, Heinrich. 1974. *Philosophies of India*. Princeton, NJ: Princeton University Press.

Zorn, Carsten. 2003. *Der Zettelkasten der Gesellschaft. Medientheorie als Gesellschaftstheorie: Eine Luhmann-Relektüre*. Dissertation. Frankfurt an der Oder: Europa-Universität Viadrina.

Index

www.ingramcontent.com/pod-product-compliance
Lightning Source LLC
LaVergne TN
LVHW102008210926
848483LV00010B/186
* 9 7 8 1 6 0 4 9 7 6 3 9 7 *